T0220092

Building Microservices Applications on Microsoft Azure

Designing, Developing, Deploying, and Monitoring

Harsh Chawla
Hemant Kathuria

Apress®

Building Microservices Applications on Microsoft Azure: Designing, Developing, Deploying, and Monitoring

Harsh Chawla
Gurugram, India

Hemant Kathuria
New Delhi, India

ISBN-13 (pbk): 978-1-4842-4827-0
https://doi.org/10.1007/978-1-4842-4828-7

ISBN-13 (electronic): 978-1-4842-4828-7

Managing Director, Apress Media LLC: Welmoed Spahr
Acquisitions Editor: Smriti Srivastava
Development Editor: Matthew Moodie
Coordinating Editor: Shrikant Vishwakarma

Cover designed by eStudioCalamar

Cover image designed by Freepik (www.freepik.com)

Distributed to the book trade worldwide by Springer Science+Business Media New York, 233 Spring Street, 6th Floor, New York, NY 10013. Phone 1-800-SPRINGER, fax (201) 348-4505, e-mail orders-ny@springer-sbm.com, or visit www.springeronline.com. Apress Media, LLC is a California LLC and the sole member (owner) is Springer Science + Business Media Finance Inc (SSBM Finance Inc). SSBM Finance Inc is a **Delaware** corporation.

For information on translations, please e-mail rights@apress.com, or visit http://www.apress.com/rights-permissions.

Apress titles may be purchased in bulk for academic, corporate, or promotional use. eBook versions and licenses are also available for most titles. For more information, reference our Print and eBook Bulk Sales web page at http://www.apress.com/bulk-sales.

Any source code or other supplementary material referenced by the author in this book is available to readers on GitHub via the book's product page, located at www.apress.com/978-1-4842-4827-0. For more detailed information, please visit http://www.apress.com/source-code.

Printed on acid-free paper

Dedicated to my son Saahir, my wife Dharna and my parents

—Harsh Chawla

Dedicated to my brother who is the guiding force in my life and career, to my parents who are always there for me!

—Hemant Kathuria

Table of Contents

About the Authors

Harsh Chawla has been part of Microsoft for last 11 years and has done various roles - currently, a Solutions Sales Professional with Microsoft GSMO. He has been working with large IT enterprises as a strategist to optimize their solutions using Microsoft technologies on both private and public cloud. He is an active community speaker and blogger on data platform technologies.

Hemant Kathuria is a consultant with Microsoft Consulting Services. He is assisting top Indian IT companies and customers in defining and adopting cloud and mobile strategies. He is an advocate of Microsoft Azure and a frequent speaker at various public platforms such as Microsoft Ignite, TechReady, Tech-Ed, Azure Conference, and Future Decoded.

About the Technical Reviewer

Devendra G. Asane is currently working as a Cloud, BigData and Microservices Architect with Persistent Systems. Prior to this he has worked with Microsoft.

Devendra lives with wife Seema and son Teerthank in Pune, India.

Acknowledgments

Harsh Chawla – I'd like to thank my wife and our families who always supported me and believed in me. Writing this book has been an enriching journey. I am eternally grateful to all my mentors especially Narinder Khullar, Pranab Mazumdar and Ashutosh Vishwakarma for their selfless support and inspiration. Lastly, thanks to the entire Apress team for their support to complete this book on time.

Hemant Kathuria – Thanks to entire Apress team especially Matthew, Shrikanth and Smriti, for helping us complete this book on time. A special thanks to Varun Karulkar for his contribution to this book.

Introduction

In the era of digital disruption, every organization is going through a major transformation. Every organization is rushing towards building businesses online. Business applications are becoming mission critical and any downtime can cause huge business impact. There is a critical need to build highly agile, scalable and resilient applications. Therefore, microservices architecture has gained huge momentum over the past few years.

This book covers the need and the key evaluation parameters of microservices architecture. It covers the scenarios where microservices architecture is preferred over the monolithic architecture, based on the learning from large-scale enterprise deployments.

The book covers practical scenarios and labs to gain hands on experience. There is an in-depth focus on the critical components for building, managing and orchestrating the microservices applications.

You will learn the following:

- Need, Evolution and Key Evaluation parameters for Microservices Architecture

- Understand the scenarios where microservices architecture is preferred over monolithic architecture

- Architecture patterns to build agile, scalable and resilient microservices applications

- Develop and Deploy Microservices using Azure Service Fabric and Azure Kubernetes Service (AKS)

- Secure microservices using Gateway Pattern
- Deployment options for Microservices on Azure stack
- Database patterns to handle complexities introduced by Microservices

CHAPTER 1

Evolution of Microservices Architecture

This is an era of digital transformation. Easy access to the Internet has empowered people and organizations to achieve more from sitting anywhere in the world. There was a time when every company was rushing for a web presence, and Internet adoption was picking up speed. Today, the Internet is in every pocket: you can access the Internet anywhere through cell phones, laptops, tablets, or PCs. This has brought a radical change to every industry. Today, every organization wants to do business online—whether retail, finance, entertainment, gaming, or so forth.

With the Internet in every pocket, the user base is increasing exponentially. A million downloads of an application or millions of views of a video within a single day are very common. Companies are running successfully in different geographic locations, and there is fierce competition to maintain market share. It's important to be highly performant and handle unpredictable user loads without interruption of service.

When millions of users are accessing a mobile application or website, even a small unhandled exception can have a cascading effect. It can bring down an entire application and cause a company to lose millions

© Harsh Chawla and Hemant Kathuria 2019
H. Chawla and H. Kathuria, *Building Microservices Applications on Microsoft Azure*,
https://doi.org/10.1007/978-1-4842-4828-7_1

of dollars. Therefore, every tiny detail in the application architecture is important. Application architectures have transformed drastically in the last ten years. Earlier, choices in technology were limited, and user load was limited as well. Application architects designed applications based on a monolithic three-tier architecture. As user loads and choices in technology increased, companies found monolithic applications difficult to scale and less agile at adopting new changes. Therefore, service-oriented architecture—and later, microservices architecture—started getting traction in the market.

This chapter covers the concepts of monolithic architecture and microservices architecture. In the coming chapters, we will delve into the technology and infrastructure details to host microservices–based applications on Microsoft Azure Cloud.

Key Evaluation Parameters

This section covers basic information about both monolithic and microservices architecture. Based on our experience, we consider four basic parameters essential to designing an application architecture. We will evaluate both monolithic and microservices architecture against these parameters.

- Scalability

- Agility

- Resilience

- Manageability

Scalability

Scalability is the capacity of an application to embrace changing user demand. It's divided into two categories.

- **Horizontal**. When the number of instances of a service or a server-like app/web tier are increased by adding more compute servers, it's called *horizontal scaling* (scale out). It's more applicable to stateless services, where there is no need to persist any state.

- **Vertical**. When the hardware capacity on the machine is increased by adding more compute (CPU/memory), it's called *vertical scaling* (scale up). It's more applicable to stateful services like databases, where data consistency must be maintained.

Agility

The flexibility of a system to embrace new changes in business functions or technology—with minimal or no interruption in the service—is called *agility*. For example, if there is a need to add another module in an HRMS application using a different technology, the system should have the flexibility to embrace this change with limited or no interruption to the entire application.

Resilience

Resilience is the ability to handle failure without interruption of service to the end user.

The following are important aspects of resilience.

- **High availability**. There are two major aspects to consider here.

  – The ability of the application to process end user requests without major downtime by maintaining that multiple instances of the same application are up and running.

  – The ability to provide limited/business-critical functionality in case of a failure at the infrastructure level or in a module of an application.

- **Disaster recovery**. The ability of the application to process user requests during the disruption caused by a disaster affecting the entire infrastructure

Manageability

It's important to understand how development operations will be managed and how easy it is to onboard new developers. This includes managing the code base of an application, change management, and the deployment process with minimal or no human intervention; for example, implementing continuous integration (CI) and continuous deployment/ delivery (CD) pipelines for efficient DevOps is an important aspect to consider.

Monolithic Architecture

Monolithic architecture is the conventional way to design an application. In monolithic architecture, the entire logic of an application is inside a single assembly or multiple assemblies deployed as a single unit. With a monolithic architecture, an application's modules are tightly coupled and are highly interdependent. Although an application can interact with other services, the core of the application runs within its own process, and the complete application is deployed as a single unit.

The application architecture is mainly divided into three tiers: presentation, business, and data access. These tiers allow functionality to be reused throughout the application.

- The **presentation tier** (user interface) is the application interface for user access. This layer allows users to perform data entry and manipulation operations.

- The **business tier** is the backbone of a monolithic application and contains the business logic, validation, and calculations related to the data.

- The **data tier** encapsulates connectivity and persistence details to access the back-end data source from higher-level components, such as the business layer. Layers also make it much easier to replace the functionality within the application. For example, an application team might decide to move a database from MySQL to SQL Server for persistence. In this case, the data tier will be impacted because changes are needed in the data tier only.

Let's take the example of an application for an HR management system (HRMS), as depicted in Figure 1-1. There are various modules, such as employee payroll, timesheets, performance management, and benefits information. In a typical monolithic architecture, all of these modules reside in a single code base hosted as a single package on web, application, and database servers.

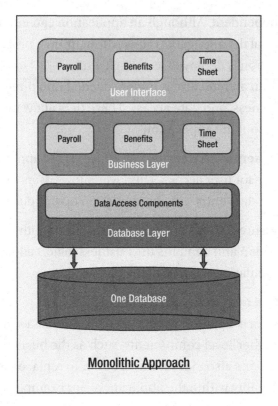

Figure 1-1. *HRMS application*

Let's evaluate the monolithic architecture based on key, defined parameters.

Scalability

Monolithic applications consist of three tiers: presentation, business, and data. Scalability can be applied at each tier.

- **Presentation tier/application tier**. These layers of an application can be designed to take the benefits of both horizontal and vertical scaling. For horizontal scaling, this layer can be designed as stateless. If it's required to design stateful services, horizontal scaling can be achieved by introducing a caching service; otherwise, vertical scaling can be achieved by increasing the compute power at the infrastructure level.

- **Data tier**. On the data tier, a hosted database stores all the data pertaining to the application. Either RDMS or NoSQL-type databases can be used to store the data. Based on the capabilities of the chosen database's technologies and business requirements, the data tier can either scale vertically or horizontally.

By design, an important factor to note about monolithic applications is that **the solution scales as a group**. Let's take an HRMS application as an example; even if you want to scale up the payroll module of the application during month/year end, the entire application has to scale up as a group.

Agility

Agility is an application's ability to embrace change in terms of technology or functionality.

- **Functionality changes**. For monolithic applications, making any change in the code needs extensive unit, stress, and integration testing efforts. Since there is a higher interdependence within the code, any major change requires testing the entire application. Even though there are options to perform automated unit testing and stress testing, any change needs lots of due diligence before rolling it out to production. With the adoption of the agile methodology for application development, application change cycles are frequent. Project teams create smaller sprints, and testing

efforts are required for each sprint to support product
rollouts every day/week/month. Managing monolithic
applications to support changes at shorter intervals
brings a lot of operational overhead.

- **Technology change**. For monolithic applications,
  making any technology-related changes or allowing
  multiple technologies in a solution is very difficult.
  Architects prefer to adopt polyglot architecture and
  need the freedom to choose different technologies
  based on business and functional needs. For example,
  a data tier document store (NoSQL DB) may be
  the preferred choice for maintaining catalogs for a
  shopping website, and an RDMS DB may be preferred
  for maintaining transactions-related data. Similarly, on
  the front end, the technology choice may be ASP.NET
  or Angular, and the API layer may be Python or
  PHP. This type of flexibility is difficult to achieve with
  monolithic applications.

Resilience

Monolithic applications are highly interdependent and hosted as a single
code base; therefore, resilience is difficult to achieve and must be carefully
designed.

Table 1-1 is a quick reference on the number of minutes that an
application can be down, based on SLA to support.

Table 1-1. *Duration That an Application Can Be Down*

SLA	Downtime/Week	Downtime/Month	Downtime/Year
99%	1.68 hours	7.2 hours	3.65 days
99.9%	10.1 minutes	43.2 minutes	8.76 hours
99.95%	5 minutes	21.6 minutes	4.38 hours
99.99%	1.01 minutes	4.32 minutes	52.56 minutes
99.999%	6 seconds	25.9 seconds	5.26 minutes

With a higher SLA requirement, there is a greater need to have an automated mechanism to detect and respond to failure. Since a monolithic application is deployed as a single code, disruption at any level can bring down the entire application.

Let's use the example of the HRMS application. If the application's database server fails (and since the database is shared by all the application's services), it can bring the entire application down. Even if high availability is in place, there will be a disruption for all the modules. Managing checks at each level of a large application becomes cumbersome.

Manageability

Manageability defines how efficiently and easily an application can be monitored and maintained to keep a system performant, secure, and running smoothly.

- **Code maintainability**. Large monolithic code bases make it difficult to onboard new developers because the code becomes very complex over time. This results in a slow feedback loop with large test suites, and it becomes cumbersome for the developers to run the full test suite locally before checking the code.

- **Monitoring**. It's much easier to monitor the code with a single code base.

Microservices Architecture

Microservices is an approach in which an application is divided into smaller sets of loosely coupled services. The purpose of a microservice is to implement a specific business functionality, and it can be easily developed and deployed independently. The microservices approach is preferred for distributed, large, and complex mission-critical applications.

Here are few principles that are essential to designing microservices.

- A microservice implements a specific business functionality.

- A microservice manages its own data and does not share databases/data models with other microservices.

- A microservice has its own code base, but there can be common components shared across different services.

- A microservice is deployed independently.

- Cross-cutting concerns like authentication should be offloaded to the gateway.

Figure 1-2 depicts the basic differences between monolithic architecture and microservices architecture.

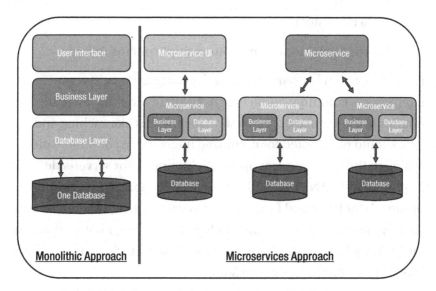

Figure 1-2. *Monolithic approach vs. microservices approach*

The small size of the service allows it to be developed and maintained by focused teams, which results in higher agility. Also, since each service is independent, you can easily adopt a polyglot architecture by making use of different programming languages/technologies to develop a service.

After the functionality is divided into multiple smaller services, interservice communication happens with well-defined interfaces using protocols like HTTP/HTTPs, WebSockets, and so forth. The most common adopted protocol is HTTP/HTTPs. In some scenarios, message queues like Azure Storage Queues and Azure Service Bus are used for higher performance and scalability.

Microservices offer many capabilities but a careful examination should be performed when designing the boundary of a microservice. It raises many new challenges related to distributed application development.

- Disparate data models

- Interprocess communication

- Data consistency

- Information monitoring from multiple microservices

Performance is a key parameter for the success of any system, but interprocess communication is a costly affair. After deciding the boundaries of the services, optimization techniques like caching and queuing should be adopted for optimized interservice communication.

Although the microservices approach is popular for server-side applications like ASP.NET Web API, many application architects adopt this approach for front-end functionality as well. The pattern can be well adopted by using technologies like Node.js, Java Spring Boot, and so forth, but the primary focus of this book is on .NET–based technologies like ASP. NET Web API on Microsoft Azure Cloud.

Let's evaluate the microservices architecture per the four key parameters.

Scalability

One of the major advantages of adopting the microservices approach is that it allows each service to scale independently as needed. Let's reference the HRMS application; a performance module will be heavily utilized during the last week of every quarter as the employees need to enter their performance goals.

Let's say that performance services have been designed as an independent microservice; as a result, it can scale up seamlessly in the last week of every quarter. Moreover, this approach can optimize the utilization of hardware resources by making resources available for other services during lean periods of the performance module. Many orchestrators also allow autoscaling based on resource needs; for example, you can trigger the scale up if the service is using more than two cores on average.

Scalability has a big advantage over monolithic architecture since most of the functionality is hosted within a few processes, and scaling requires cloning the complete app on multiple servers. Although cloud providers like Azure offer great support for scaling monolithic apps by using services like the Azure App Service, the microservice approach is more efficient because it deploys and scales each service independently across multiple servers.

Agility

Each microservice is an independent subsystem, and communication among these subsystems happens over common protocols. This simplifies functionality changes, and it's much easier to choose different technologies for development.

Functionality Changes

Since the entire application is divided into multiple independent services, adding any change to a service doesn't impact the entire solution. This allows you to support quick bug fixes and shorter life-cycle releases. The microservice approach really compliments the agile style of software development.

With monolithic applications, a bug found in one part of an application can block the entire release process. As a result, new features may be delayed due to a bug fix that is pending to be integrated, tested, and published.

Adding a new module is much simpler and can be easily done without interrupting the existing application.

Technology Change

As per the design of microservices architecture, coding a service module can be done in a completely different programming language. Let's say there are two services in an HRMS application.

- Employee performance service

- Employee demographic service

These services communicate based on well-defined exposed interfaces and common communication protocols; therefore, each service can be developed in a different technology—one in .NET and one in Java. Although it may involve certain complications, a cautious approach can be adopted to avoid vendor lock-in and harness the best technology stack to build the application.

Resilience

Since each microservice is an independent subsystem and has its own database, the unavailability of one service does not impact another service. Also, the system can be made available during a service's deployment/updates. This is one of the major advantages over monolithic applications—any updates to a monolithic application make the entire system unavailable, which requires longer downtimes.

In most scenarios, microservices are managed by orchestrators that make sure that the microservices are resilient to failures and able to restart on another machine to maintain availability. If there is a failure during the deployment of microservices, orchestrators can be configured to continue to the newer version, or rollback to a previous version to maintain a consistent state.

Manageability

The manageability of microservices is covered in this section.

Code Maintainability

Since each microservice is an independent subsystem and focused on one specific business functionality, it is easy for development teams to understand functionality. This really helps when onboarding new developers during the later stages of a project. Also, microservices compliment the agile methodology, where application development efforts are divided across teams that are smaller and work more independently.

Monitoring

Monitoring multiple microservices is more difficult because a correlation is required across services communication.

Comparison Summary

Table 1-2 is a summary of monolithic architecture and microservices architecture, based on the key parameters defined.

Table 1-2. *Comparison Summary*

Key Parameter	Monolithic Application	Microservices
Scalability	The application scales as a group.	Allows each service to scale independently based on the need of the resources.
Agility	Making any change in the code and functionality needs extensive testing as the modules are tightly coupled and are deployed as single unit.	Entire application is divided into multiple independent services, adding any change to a service doesn't impact entire application. This allows supporting quick bug fixes and shorter life-cycle releases.
Resilience	The application is highly interdependent and hosted as a single code base. Therefore, resilience is difficult to achieve and must be carefully designed.	The entire application is divided into multiple independent services and has its own database; the unavailability of one service does not impact another service. Also, it makes easy to make the system available during the deployment/updates of a service.
Manageability	Code maintainability is difficult because large code bases make it difficult to onboard new developers. Monitoring is easy because it's a single code base.	Code maintainability is easy because multiple independent services allow development teams to understand the functionality quickly. Monitoring is a challenge because a correlation needs to be managed across multiple services.

Challenges of Microservices

Microservices is a highly adopted architecture and the most popular approach for designing large-scale, complex systems. However, there are certain areas that need to be looked at carefully and mitigated before finalizing the microservices approach for your application.

Database Redesign

Even though microservices is a popular approach, enterprises find it difficult to re-architect their databases and schemas. One of the fundamental principles of microservices is that a service should own its data. It should not depend upon a large, shared, central repository. If a service relies upon a system of record—like mainframes, SAP systems, and others, then the design of microservices do not adhere to the definition. The following are the most common challenges faced when designing a database.

- Sharing or making the master database records available across microservices databases.

- Maintaining foreign keys and data consistency when the master records are available in a common database.

- Making data available to reports that need data from multiple microservice databases.

- Allowing searches that need data from multiple microservice databases.

- If the creation of a record requires multiple microservices calls, making sure that the database is consistent, because both calls cannot execute in a single database transaction. Atomic transactions between multiple microservices are usually a challenge, and business requirements must accept eventual consistency between multiple microservices as a solution.

Interservice Communication

Interservice communication is one of the most common challenges faced when using the microservices approach. There are a few challenges under this category, which are highlighted as follows.

- Microservices communicate via well-defined interfaces and protocols like HTTP/HTTPs, which adds **to the latency of the system** when a business operation requires multiple microservice calls. Due to the involvement of multiple microservices for the completion of a single business operation, this makes testing difficult and increases the testing team's overhead.

- Another challenge that crops up is whether the client application should invoke multiple microservices, or a gateway/facade service should be introduced to invoke child microservices.

- A client invoking multiple microservices can be a challenge when the client is on a mobile network. Due to the limited bandwidth, invoking multiple services for the completion of a single business operation is not efficient.

- Security is another important aspect of the microservices architecture. It's critical to decide whether each microservice should be responsible for its own security, or if a gateway/façade service should be introduced to maintain the security of child services.

Higher Initial Expense

Since each service runs its own process and maintains its own database and schema, initial expenses can be much higher. Each service needs separate compute and storage, which increases the footprint of the application. Therefore, containers are highly adopted in microservices architectures. A container is much smaller than a virtual machine, and it can help optimize costs.

Deployment Complexities

One of the major advantages of a microservices approach is that you can scale up a service based on the load; however, the deployment becomes complex and increases complexity of IT operations. There are many orchestrators—like Azure Service Fabric (it has many more features than an orchestrator) and Azure Kubernetes Service (AKS)—to ease out the deployment efforts. However, learning how to manage these orchestrators will be an added responsibility for a team.

Monitoring

Monitoring multiple microservices is lot harder than a monolithic application. Monitoring business operations that span across multiple services needs a lot of correlations to identify the issues. Platforms like Azure Service Fabric have good, built-in health indicators to monitor services. Proper planning and design are required for it to work seamlessly.

Versioning

Versioning is the most critical part of any application. Releasing a new version of a service should not break the other dependent services. This factor should be carefully planned to support backward and forward compatibility. Moreover, if a failure occurs, there should be a provision to automatically rollback to the previous version. All such issues can be easily solved with orchestrators like Azure Service Fabric or Azure Kubernetes Service.

Summary

When designing a solution, the most important question is how to architect the solution. In this chapter, we shared enough information to help you understand the benefits of Micro Services Architecture over Monolithic Applications. Since the focus of this book is microservices, we focus on how a microservices architecture can be implemented and how the power of cloud computing (i.e., Microsoft Azure) can be harnessed to build a cost-effective and highly robust application.

CHAPTER 2

Implementing Microservices

Microservices is a preferred way of architecting large-scale and mission-critical applications. It offers many advantages, such as using multiple technologies in the same solution; for example, NoSQL is preferred for building shopping website catalogs, and RDMS can manage transactions in the back end. Similarly, one service can be developed in ASP.NET and another in Java. In the past, application developers had to work in single development language, and developers specializing in a different technology had to be trained to make them relevant to the project. Microservices offers the freedom to use the best technology for a project and to pick experts from various technologies.

While the microservices architecture has lots of advantages in terms of scalability, resilience, agility, and manageability, it brings challenges as well. Let's delve deeper into the critical factors to consider when building a microservices ecosystem.

- Client-to-microservices communication

- Interservice communication

- Data considerations

- Security

© Harsh Chawla and Hemant Kathuria 2019
H. Chawla and H. Kathuria, *Building Microservices Applications on Microsoft Azure*,
https://doi.org/10.1007/978-1-4842-4828-7_2

- Monitoring

- Hosting-platform options

- Choice of Orchestrator

- Orchestrator

Client-to-Microservices Communication

One of the challenges of the microservices architecture is client-to-microservices communication. In a microservices approach, an application is divided into smaller sets of loosely coupled services, and the boundaries of microservices are defined based on a decoupled applications domain model, data, and context. The real challenge is how to retrieve data for a business scenario that involves invoking multiple microservices. An API gateway is one of the popular solutions for handling this challenge.

API Gateway

In a microservices architecture, each microservice exposes a set of fine-grained endpoints. Typically, due to the nature of client applications, data needed by a client involves the aggregation of data from different microservices. Since a screen in a client application may require data from multiple microservices, a client application can connect to multiple microservices individually, which may lead to certain issues.

- If a client app is a server-side web application like ASP.NET MVC, the communication to multiple microservices can be efficient. With mobile and SPA clients, chatty communication to multiple microservices can cause overhead due to network connectivity and the performance of the mobile devices.

- Due to the nature of client applications, the data needs of clients can be very different; for example, a desktop-based version of an employee information page may show more than the mobile-based version.

- Each microservice is responsible for cross-cutting concerns like authentication, authorization, logging, and throttling, which are major overhead from a design and development perspective.

When looking at these challenges, it can become a nightmare for an architect to manage a client application invoking multiple microservices. One of the best possible solutions to this problem is to implement an API gateway solution.

At a higher level, an API gateway is an entry point service for a group of microservices. It is very similar to the facade pattern in object-oriented design. This pattern reduces chattiness between the client and the services.

It is important to highlight that having a single custom API gateway service can be a risk if not implemented correctly. The gateway service grows and evolves based on the requirements of the client apps. Eventually, it can lead to a scenario very similar to a monolithic application. Hence, it is recommended to split the API gateway between service categories; for example, one per client app, form factor type, and so forth. API gateways should never act as a single aggregator for all internal microservices. It should be segregated based on business boundaries.

API gateways can provide different functions and features, which can be grouped into the following design patterns.

Gateway Aggregation

As depicted in Figure 2-1, this pattern reduces chattiness between the client and the services. The gateway services become the entry point. They receive client requests and hand over these requests to the various back-end microservices. The gateway also **aggregates and combines the results** and sends them back to the requesting client. This pattern performs well when a client needs to interact with multiple services for a single business scenario and data aggregation is needed from multiple services. It also helps in scenarios where the client is operating on a high-latency network, such as mobile phones.

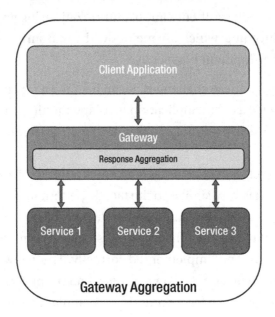

Figure 2-1. *Gateway aggregation*

The following considerations should be applied when implementing this pattern.

- A gateway service should be located near the back-end microservices to reduce network latency. A gateway service should not interact with services located across data centers.

- A gateway should not become a bottleneck. It should have the ability to scale on its own to support the application load.

- Scenarios in which one of the services times out and partial data is returned to the client application should be handled carefully.

- Performance testing should be done to make sure that the gateway service is not introducing significant delays.

- Distributed tracing should be enabled with the help of correlation IDs to enable monitoring in case of failure and for diagnosis.

Gateway Routing

As depicted in Figure 2-2, this pattern is very similar to gateway aggregation, where, primarily, the gateway only routes the requests to multiple services using a single endpoint. This pattern is useful when you wish to expose multiple services on a single endpoint and route to the appropriate service based on the client request. In a scenario where a service is discontinued or merged with another service, the client can work seamlessly without an update as the intelligence of routing is handled by the gateway and changes are required only at the gateway level.

In an enterprise scenario, one use case for gateway routing is to expose on-premise APIs to the outer world on the Internet. In a scenario where you have to expose an API to a partner, for a vendor that is not connected to a corporate network, an API gateway can expose a public endpoint with the required security and can internally route the traffic to on-premise APIs.

Azure Application Gateway is a popular managed load-balancing service that can implement a gateway routing pattern.

To support this pattern, layer 7 routing is used by the gateway to route the request to the appropriate service. Since gateways introduce a layer of abstraction, a deployment and service update can be easily handled. It also allows supporting multiple versions of a service by introducing a new route or by routing the traffic internally to an older or a newer service endpoint without affecting the client.

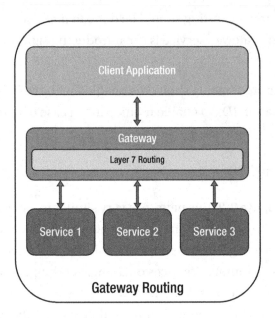

Figure 2-2. *Gateway routing*

The following considerations should be made while implementing this pattern.

- The gateway service should be located near the back-end microservices to reduce network latency. The gateway service should never interact with services located across data centers.

- A gateway should never become a bottleneck. It should have the ability to scale on its own to support the application load.

- Performance testing should be done to make sure that the gateway service is not introducing significant delays.

- A gateway route should support layer 7 routing. It can be based on the IP, port, header, or URL.

Gateway Offloading

As depicted in Figure 2-3, the gateway offloading pattern helps offload the cross-cutting concerns from individual microservices to the gateway service. It simplifies the implementation of each microservice as it consolidates cross-cutting concerns into one tier. With the help of offloading, specialized features can be implemented by a specialized team and at one common tier. As a result, it can be utilized by every microservice.

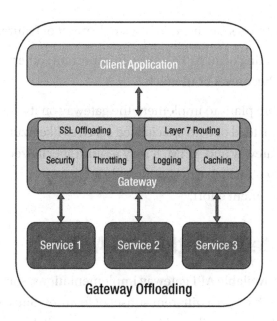

Figure 2-3. *Gateway offloading*

Most of the common cross-cutting concerns that can be effectively handled by the API gateway are

- Authentication and authorization

- Logging

- Throttling

- Service discovery

- SSL offloading

- Response caching

- Retry

- Load balancing

The following considerations should be made while implementing this pattern.

- API gateways can introduce a single point of failure.

- The scaling of an API gateway is important; otherwise, it can become a bottleneck.

- If a team plans to implement the gateway on its own, instead of with specialized services like Azure API Gateway and APIM, it may require specialized resources, which can significantly increase development effort.

The API Gateway Pattern on Azure

There are many available API gateway implementations, such as Kong and Mulesoft, and each offers a different subset of features. Since the focus of this book is on Azure, we will explain the options available on Azure.

- **Azure Application Gateway**. Application Gateway is a managed load-balancing service that performs layer 7 routing and SSL termination. It also provides a web application firewall (WAF).

- **Azure API Management**. API Management offers publishing APIs to external and internal customers. It provides features such as rate limiting, IP white listing, and authentication using Azure Active Directory or other identity providers. Since API Management doesn't perform any load balancing, it should be used in conjunction with a load balancer such as Application Gateway or a reverse proxy.

Interservice Communication

With monolithic applications, a component can easily invoke another component as they are running in the same process. Also, the language-level constructs (like new `classname()`) can be used to invoke methods on another component. One of the challenges with microservices architecture is handling interservice communication as the in-process calls change to remote procedure calls.

Also, if a microservice is invoking another microservice heavily, it defeats the basic principal of microservices. A fundamental principal of the microservices architecture is that each microservice is autonomous and available to the client, even if the other services are down or unhealthy.

There are multiple solutions to this problem. One solution is to carefully decide the boundary of each microservice. This allows the microservice to be isolated, autonomous, and independent of other microservices. Communication between the internal microservices should be minimal.

If communication is required, asynchronous communication should take priority over synchronous communication because it reduces coupling between services. It also increases responsiveness and multiple subscribers can subscribe to the same event. In asynchronous messaging, a microservice communicates with another microservice by exchanging messages asynchronously. If a return response is expected, it comes as a different message and the client assumes that the reply will not be received immediately, or there may not be a response at all.

Asynchronous messaging and event-driven communication are critical when propagating changes across multiple microservices, and they are required to achieve eventual consistency, as depicted in Figure 2-4.

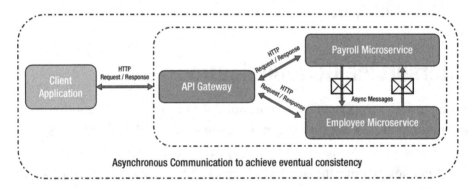

Figure 2-4. *Asynchronous communication to achieve eventual consistency*

Asynchronous messages are usually based on asynchronous protocols like AMQP. Message brokers are generally preferred for these kinds of communications (e.g., RabbitMQ, NServiceBus, MassTransit) or a scalable service bus in the cloud, like Azure Service Bus.

If there is a need to **query real-time** data (e.g., to update the UI), generally, **request/response communication with HTTP and REST** is used to support these kinds of scenarios. In this kind of pattern, the client assumes that the response will arrive in a short time. If synchronous communication is required between services, you can take advantage of binary format communication mechanisms (e.g., Service Fabric remoting or WCF using TCP and a binary format). You can also take advantage of caching using the cache-aside pattern. You should be careful of adopting this pattern because having HTTP dependencies between microservices makes them non-autonomous and performance is impacted as soon as one of the services in the chain does not perform well. This architecture can be easily designed and developed by using technologies such as ASP.NET Core Web API, as depicted in Figure 2-5.

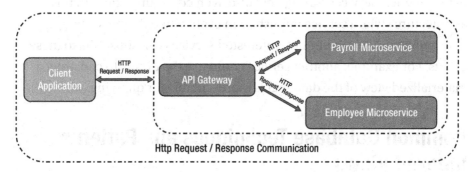

Figure 2-5. *HTTP request / response communication*

Data Considerations

A basic principle of microservices is that each service should manage its own data. Each service should be responsible for its own data store, and other services should not access it directly. This prevents coupling between services and allows the services to scale based on load needs. This principle also allows the services to use different database technologies, if required.

Due to the distributed approach of managing data, there are certain challenges that occur, like redundancy of data across data stores. One of the major challenges is propagating the updates across services, because it is not possible to spawn a database transaction across multiple services.

31

There are multiple solutions to this problem. One solution is to embrace eventual consistency wherever possible; we should clearly distinguish use cases where ACID transactions are required and where eventual consistency is acceptable. In scenarios where strong consistency is required, one service should represent the source of truth for a given entity. In scenarios where transactions are needed, patterns such as compensating transactions can be used to keep data consistent across several services.

Finally, a good solution for this problem is to use eventual consistency between microservices through event-driven communication. In this architecture style, a service publishes an event when there are changes to its public models or entities. Interested services can subscribe to these events. For example, another service could use the events to construct a materialized view of the data that is more suitable for querying.

Common Database Techniques and Patterns Indexed Views

Let's say that we have an HRMS system client application that needs to display employee's personal information along with payroll details, and there are two microservices involved (i.e., employee and payroll services). As per the basic principal, each microservice owns its data and the application reads and writes the data only via well-defined interfaces. Since employee information will be needed on almost all the screens of the client application, a denormalized read-only table can be created to be used only for queries. Since the view is created in advance, and it contains denormalized data, it supports efficient querying. An important point to note is that the data in the indexed view is completely disposable because it can be entirely rebuilt from the source databases.

Another classic use case that can be efficiently handled by this approach is replicating the master data that is required by almost all the microservices. Having an HTTP call across the services or across the database joins can be an inefficient approach from the performance and dependency perspective; hence, indexed views effectively solve this problem.

Here are a few ideal reasons for using this technique.

- Indexed views significantly improve query performance for reporting and display needs.

- In cases where the source data is available in normalized form and require complex queries, having an indexed view removes complexity while reading the data.

- It allows access to data based on privacy needs.

- It effectively supports disconnected scenarios, in which the source database is not always available.

Please note that this technique can be inefficient if changes to the source database are frequent and source data accuracy is a priority.

Data Warehouse for Reporting Needs

To support complex reports and queries that don't require real-time data, a common approach is to export data into large databases. That central database system can be a big data–based system, like Hadoop, a data warehouse, or even a single SQL database only used for reports.

An in-depth discussion of this subject is in Chapter 8.

Security

Since a microservices application is distributed in multiple services, client authentication and authorization becomes challenging.

A commonly suggested practice for handling security centrally is to use an API gateway. In this approach, the individual microservices cannot be reached directly, and traffic is redirected to an individual API via a gateway once a successful authentication is performed.

On Microsoft Azure, the API gateway service is readily available and provides features like authentication, IP filtering, and SSL termination, and helps avoid exposing microservice endpoints directly.

Ocelot develops a custom API gateway. It is an open source, simple, lightweight, .NET core–based API gateway that can be deployed for microservices. For authorization, Azure Active Directory helps manage role-based access for the microservices' resources.

Monitoring

Monitoring is an important aspect in understanding how microservices are performing in a deployed environment.

On Microsoft Azure, services like Application Insights and Azure Monitor are used for monitoring, logging and alerting on both Azure and on-premise resources. Azure Monitor helps you draw insights on API access patterns and identify the root cause of any performance issues. It can also work in conjunction with Application Insights to draw insights on the application's performance.

Microservices Hosting Platform Options

An important area of discussion is whether to host microservices on virtual machines or on containers. Both options can be implemented on-premise and on cloud platforms like Microsoft Azure. On Azure, the management of an underlying infrastructure becomes very easy because there are specialized services available, such as Azure Serve Fabric and Azure Kubernetes Service (AKS). Using containers to implement microservices is the most preferred option, and it's important to understand the various reasons behind it. In this section, we explain both hosting platform options.

Using Virtual Machines

A virtual machine is an operating system installation on the virtualization layer of the physical host, as depicted in Figure 2-6. It helps to optimize the hardware utilization by enabling the physical host to provide an isolated environment for each application. The caveat is that for every virtual machine, an entire OS must be installed separately.

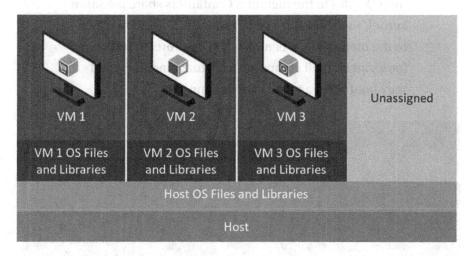

Figure 2-6. *Virtual machine hosting*

Therefore, every virtual machine needs to boot up and load OS files into its memory. This mechanism dissipates lots of compute resources on the host operating system.

Using a Container

Containers are like virtual machines. They offer a way to wrap an application into its own isolated box using namespace isolation. In this technique, the host OS creates a namespace for all the resources (e.g., disk, memory, running process, etc.) to make the environment look as if dedicated for the container.

Containers differ from virtual machines in a few ways.

- Virtual machines have a complete OS installation on the virtualization layer of the physical host. It takes time to start up because it must boot the entire OS and map OS files in the memory. Containers share the same kernel, so there is no need to boot the OS and map files to the memory (see Figure 2-7). Therefore, a container footprint is small compared to virtual machines, and they boot up in a much shorter time.

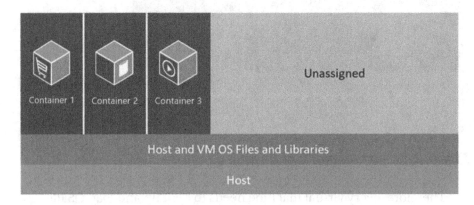

Figure 2-7. *Containers sharing OS files and libraries*

Starting with Windows 2016, there is an option to host containers in Hyper-V mode, which separates the kernel of a container from the host OS. It can be used for highly sensitive applications or for multitenant environments. Start-up efficiency reduces when compared to containers that use namespace isolation, however.

- Since containers make the environment and resource consumption consistent, it becomes convenient for developers to run the same application on different systems without change in experience. With virtual machines, however, applications can perform differently in different environments.

Microservices segregate an entire solution into multiple services; agility, scalability, and optimum resource utilization are the most important factors. Since containers perform much better than VMs for such scenarios, they are an enterprise's first choice for a deployment platform.

Let's look at the basic components of a container ecosystem.

- Container image
- Container registry
- Container

Container Image

A container image is like a software installer file that contains both the OS layer and the application layer, with all the dependencies to run the application (e.g., Windows Nano Server, SQL Server). A container image can be used numerous times to install an application.

Container Registry

The container registry is the repository for the container images that can be made accessible to the entire organization. Any authorized user in an organization can push or pull images from this repository. It can be created

as either public or private, depending on the requirements. Microsoft
Container Registry is a public registry that hosts images for public
download. Azure Container Registry is used for maintaining a private
registry.

The command to download an image from a public repository is

```
docker pull mcr.microsoft.com/mssql/server:2017-latest
```

Let's break down this command.

- `mcr.microsoft.com/mssql/server:2017-latest` is the container image.

- `mcr.microsoft.com/mssql/server` is the container registry.

- `2017-latest` is the tag.

- `docker pull` – `docker` is the command line to pull an image from the registry.

The local repository looks like the following.

```
REPOSITORY                        TAG                IMAGE ID       CREATED        SIZE
mcr.microsoft.com/mssql/server    2017-latest        314918ddaedf   8 weeks ago    1.35GB
mcr.microsoft.com/mssql/server    2019-CTP2.2-ubuntu 0c6e117b2c2e   2 months ago   2.02GB
docker4w/nsenter-dockerd          latest             2f1c802f322f   4 months ago   187kB
```

Container

The container is an instance of a container image. Multiple container
instances can be spun from a single container image. If a SQL Server
container is spun from the image of the local repository, it will create an
instance of SQL Server on an Ubuntu server. The experience is like a virtual
machine, where you can get in the OS layer, run commands, and work with
SQL Server from both inside and outside of the containers.

Choice of Orchestrators play a key role in managing a large number
of containers or virtual machines. High availability, scalability, and

application agility are the most critical factors that orchestrators are expected to cover. The following are the most important functionalities that an orchestrator should cover.

- **Clustering of resources**. This feature makes groups of VMs or physical machines look like a single resource. All the resources are provided from a single group. This helps optimize resource utilization, and even management becomes easy.

- **Orchestration**. This feature helps make all the components work together as a unit. Running containers, their scalability, load balancing during a heavy load, and high availability during failures are ensured by this functionality.

- **Management**. Managing networking, storage, and services come under management functionalities of the orchestration tools.

Here are a few of the orchestration solutions available on the market.

- Azure Service Fabric
- Azure Kubernetes Service
- Docker Swarm and Compose
- Mesos DC/OS

Let's look at an overview of these solutions.

Azure Service Fabric

Azure Service Fabric is an orchestration tool that can be deployed both on-premises and on Microsoft Azure. This Microsoft solution manages multiple services on Azure. Applications are deployed in the form of services on Azure Service Fabric. Every service (stateless or stateful) has three components.

- Code

- Configuration

- Data

A Service Fabric cluster is built on a bunch of physical or virtual machines called *nodes*. There are various services (e.g., failover manager services, repair manager services, naming services, etc.) to manage high availability, health, and service identification for Azure Service Fabric. Apart from containers, services can be run as guest executables and reliable services by using the native Service Fabric SDK.

Azure Kubernetes Service

Kubernetes is an orchestration tool that can be deployed both on-premise and on Microsoft Azure. On Azure, it's a managed service named Azure Kubernetes Service. On AKS, pods run a single or a group of containers. Services are the labels that point to multiple pods. Kubernetes has a cluster master and cluster nodes to manage a container ecosystem.

Docker Swarm and Compose

Docker Swarm is the clustering tool for Docker; it can be deployed both on-premise and on Microsoft Azure. Each node in the cluster runs as swarm agent, and one of the nodes run a swarm manager. A swarm manager is responsible for orchestrating and managing the containers on the available container's host. Filters can be configured on Docker Swarm to control the hosting of containers. Docker Compose is a command-based utility to configure an application's services. With a single command, an entire application can be up and running on the swarm cluster.

Mesos DC/OS

The Apache Mesos orchestration solution is designed to handle a large number of hosts to support diverse workloads. It can be run both on-premise and on Microsoft Azure. This setup has a Mesos master to orchestrate the tasks, and agent nodes to perform the tasks. Frameworks coordinates with the master and schedules tasks on agent nodes.

Summary

In this chapter, we explained how to overcome the challenges introduced by adopting the microservices architecture. We listed some of the platform-hosting options for microservices. In the coming chapters, we discuss these options in more detail.

CHAPTER 3

Azure Service Fabric

In the previous chapters, you learned about the evolution of microservices, the advantages and challenges of the microservices architecture. We also described various options for implementing microservices in Azure. In this chapter, we will explain Azure Service Fabric's offerings.

Although Service Fabric is a vast subject and needs a complete book explaining its concepts, fundamentals, and various flavors. In this chapter, we cover the fundamentals and real-life experiences that you should be aware of. Hands-on examples give you a better idea of how easy it is to adopt Service Fabric for implementing a microservices architecture.

What Is Service Fabric?

Azure Service Fabric is a distributed systems platform that allows you to run, manage, and scale applications in a cluster of nodes, using any OS and any cloud.

The Service Fabric SDK allows you to implement service communication, scale, and service discovery patterns effectively. The SDK is available for .NET and Java developers. Please note that *it's not mandatory to use SDK*. You can develop an application in any programming language, and you can deploy over Service Fabric using guest executables and containers.

Service Fabric can be deployed on the platform of your choice (i.e., Windows or Linux) and can be deployed on-premise, on Azure or AWS, or on any other cloud platform.

© Harsh Chawla and Hemant Kathuria 2019
H. Chawla and H. Kathuria, *Building Microservices Applications on Microsoft Azure*,
https://doi.org/10.1007/978-1-4842-4828-7_3

In Service Fabric, an application is a collection of multiple services, and each service has a defined function to perform. A service is represented by three components: code (binaries and executables), configuration, and data (static data to be consumed by service). Each component is versioned and can be upgraded independently. This is one of the significant advantages of Service Fabric—in a deployment failure, you can easily roll back to any previous version of the service.

Also note that if even a simple change to an application's configuration is made, a deployment and version upgrade are required. In our experience, this is initially a problem for deployment teams because most of enterprises have stringent deployment processes, but it really helps in the event of a deployment failures because you can roll back to any of the previous versions by using a single command.

Deployment processes can be streamlined by adopting services like Azure DevOps.

Service Fabric also supports autoscaling. The autoscale feature allows Service Fabric to dynamically scale your services based on indicators like load, resources usage, and so forth.

Note   If the Service Fabric clusters are not running in Azure, the process of scaling is different, and you must manually change the number of nodes in the cluster, which is followed by a cluster configuration upgrade.

Core Concepts

In this section, we explain core Service Fabric concepts, such as the application model, scaling techniques, supported programming models, and types of Service Fabric clusters. Also, we cover certain hands-on labs to make sure that you get real-life exposure, so that you are not limited to theoretical knowledge.

Service Fabric Application Model

In Service Fabric, an application is a collection of services in which each service and application is defined using a manifest file.

Each service in an application is represented by a service package, and the package has three components (code, configuration, and data), as depicted in Figure 3-1. The code component contains the actual executables, binaries of the service, or pointers to the container images in container repositories such as ACR and Docker Hub. The configuration component contains the configuration entries required by the service; it's very similar to web.config in ASP.NET applications, and if needed, you can have multiple configuration files. The data component contains static data to be consumed by the service. Service Fabric is not very particular about the data format; it can be JSON, XML files, and so forth.

Please note that each component is versioned and can be upgraded independently. Also, the service package is always deployed as a group, which means that if you want to make two containers run together on the same node, you can include two code packages (pointing to the respective containers) in the same service package.

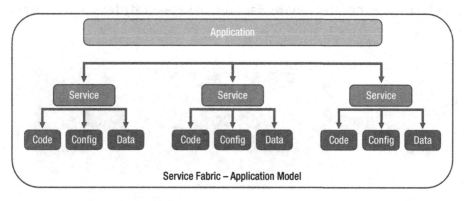

Figure 3-1. *Service Fabric application model*

While deploying an application on Service Fabric, an Application Type gets created to represent an application and a Service Type gets created to represent each service in a Service Fabric cluster. Also, on successful deployment an instance of Application Type and Service type gets created. You can have many instances of an Application type to support different version of the same application and can have multiple instances of service type to support higher load and high availability.

In my experience, I have noticed stateless services are widely used and the following techniques are mostly used to support scale and higher availability.

Scale by Increasing or Decreasing Stateless Service Instances

When creating a service instance, you can specify the instance count, which defines the number of Service Fabric cluster nodes in which your service is hosted. The following command specifies the count to two, which means that the service is hosted on only two nodes, even if the number of nodes in the Service Fabric is greater than two (see Figure 3-2).

```
sfctl service create --name fabric:/a/s1 --stateless -instance-
count 2
```

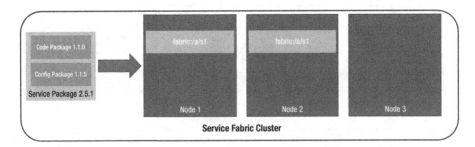

Figure 3-2. *Service Fabric cluster: two nodes utilized*

Service Fabric also allows you to update the instance count. You can set the count to –1 to instruct Service Fabric to run the service on all available nodes, as depicted in Figure 3-3. If new nodes are added to the cluster, Service Fabric makes sure that your service is hosted on the newly added nodes too.

```
sfctl service update --service-id a/s1 --stateless -instance-
count -1
```

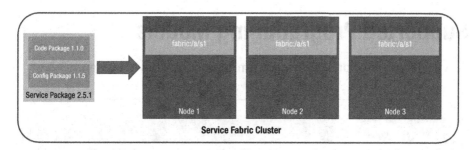

Figure 3-3. *Service Fabric cluster: all nodes utilized*

Scale by Adding or Removing Named Services Instances

In scenarios where the node capacity is underutilized, you can instruct Service Fabric to scale up by creating another named service instance and deploy the exact same code package on all the available nodes, but with different unique names (see Figure 3-4).

```
sfctl service update --service-id a/s2 --stateless -instance-
count -1
```

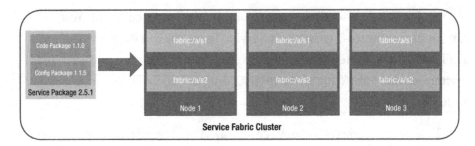

Figure 3-4. *Service Fabric cluster: named service instances*

Supported Programming Models

Service Fabric supports the programming models shown in Figure 3-5.

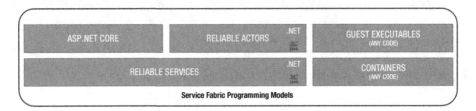

Figure 3-5. *Service Fabric cluster: supported programming models*

Containers

One of the programming models offered by Service Fabric allows you to orchestrate and deploy applications using both Windows and Linux containers. Service Fabric supports the deployment of docker containers on Linux and Windows server containers (including Hyper-V isolation) on Windows Server 2016.

Service Fabric has the capability to pull the container images from container repositories like Docker HUB and Azure Container Registry.

Deploying an application as a container does not require any changes to your application and has no Service Fabric SDK dependency.

Although you can deploy your services using multiple programming models (guest executables, stateless or stateful services), there are certain scenarios where containers are more suitable.

Monolithic Applications

If a monolithic application is developed using ASP.NET web forms and has dependency on technologies like IIS, you can package these applications as container images and deploy on Service Fabric for effective scaling and deployment management.

In this mode, you have no dependency on Service Fabric SDKs; you can deploy an application as it is. An application can also be developed in any programming language.

Application Isolation

If a complete or higher level of isolation from other applications running on the same host is required, containers are a very viable option because they provide isolation effectively. Also, Windows Containers Hyper-V mode takes isolation to a different level because the base OS kernel is not shared between containers.

Service Fabric also provides resource governance capabilities to restrict the resources that can be used by a service on a host.

Reliable Services

Reliable services allow you to write services using the Service Fabric SDK framework. It is a lightweight framework that allows Service Fabric to manage the life cycle of your services. It also allows the services to interact with the Service Fabric runtime. With SDK, you benefit from features such as notifications on code or configuration changes, and communicating with other services.

Please note that both C# developers and Java developers can develop reliable services using the Service Fabric SDK for Linux.

Note   At the time of writing this book, if you plan to use Java SDK, then you need to use a Mac or Linux developer machine.

Reliable services can be stateless or stateful. In stateful services, the state is persisted using reliable collections.

Stateful Reliable Service

A stateful reliable service allows you to store data within the service itself. Service Fabric makes sure that the state is highly available and persistent. There is no need to store the state in external storage. Stateful services uses reliable collections, which falls under the Microsoft.ServiceFabric.Data. Collections namespace. Reliable collections are very similar to collections in the System.Collection namespace, but with the following differences.

- Data is not only persisted into the memory, it's also persisted to disks to face large-scale outages.

- It supports transactions.

- State changes are replicated using replicas to support high availability.

- Reliable collection APIs are asynchronous to make sure that threads are not blocked.

Although, you have the choice of using other technologies (e.g., Redis Cache or Azure Table service to store the state externally), reliable collections make all the reads local, which results in high throughput and low latency.

These are the collections available under the Microsoft.ServiceFabric. Data.Collections namespace:

- Reliable dictionary

- Reliable queue

- Reliable concurrent queue

Each stateful service has a state associated with it. High availability is achieved with the help of replicas. Each replica is a running instance of the service code. The replica also has a state. The r/w operations are performed on one replica, which is called a *primary replica*. Changes to the state are replicated from the primary replica to other replicas, called *active secondary replicas*. In a primary replica failure, Service Fabric makes one of the active secondary replicas the primary replica.

To support high scalability, stateful reliable services use partitioning. A partition is a set of replicas responsible for a portion of the complete state of the service.

Let's say that you have a five-node cluster with five partitions, and each partition has three replicas (as depicted in Figure 3-6). In this case, Service Fabric will distribute the replicas across the nodes, and each node will end up having two primary replicas per node.

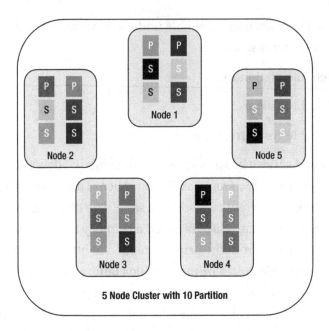

Figure 3-6. *Service Fabric: five-node cluster*

Service Fabric provides three types of partitioning schemes: ranged partitioning (UniformInt64Partition), named partitioning, and singleton partitioning. In ranged partitioning, the number of partitions and an integer range is specified for a partition with the help of a low key and a high key. By default, a Visual Studio template uses range partitioning as a default, and it is the most useful and common one.

For information on transaction and lock modes, backup, and restore, you can refer to the online documentation.

Stateless Reliable Service

Stateless reliable services do not maintain any state across the service calls. It is a familiar paradigm in web development in which you have a service layer that receives a request and connects to an external data store (e.g., Azure SQL , Azure Storage, or Document DB) to return the response.

Since there is no state maintained by the service, it does not require any persistence or synchronization.

Guest Executable

Service Fabric runs any type of code developed in any language (e.g., Node.js, Java, C++, etc.). Service Fabric considers guest executables to be a stateless service. If an application is self-contained, you can run it as a guest executable on the Service Fabric platform. Running an application as a guest executable allows the application to use the Service Fabric REST API to call the Service Fabric naming service to find other services in the cluster.

The following subsections discuss the significant benefits of hosting applications as guest executables.

Scalability and High Availability

Service Fabric allows you to dynamically scale your services based on indicators like load, resources usage, and so forth. It makes sure that the instances of your application are running, and in the event of any failure, reports the same on the Service Fabric dashboard.

Health Monitoring

With Service Fabric dashboards, you can determine the health of an application. It provides diagnostics information in the event of a failure.

Application Deployment

You can deploy the applications with no downtimes. You can benefit from features like automatic rollbacks in the event of a deployment failure.

Optimizing Hardware

This allows you to run multiple guest executables in the same cluster; hence, you are not required to maintain a set of hardware for each application.

ASP.NET Core

ASP.NET Core is a new, open source, cross-platform framework for building modern, cloud-based, Internet-connected, applications, such as web apps, IoT apps, and mobile back ends. ASP.NET Core–based applications can be hosted as guest executables or as a reliable service, in which it takes advantage of the Service Fabric runtime.

Reliable Actors

The reliable actor framework is based on reliable services. It implements the actor design pattern.

Service Fabric Clusters

Service Fabric helps you deploy your microservices on a cluster, which is a set of virtual or physical machines that are interconnected through a network. Each machine inside the cluster is called a *node*. A cluster can consist of thousands of nodes, depending on resource needs of your application. Each node in a cluster has a Windows service called FabricHost.exe, which makes sure that the other two executables (Fabric. exe and FabricGateway.exe) are always running on the cluster nodes.

Service Fabric cluster can be created using virtual machines or physical machines running on Windows or Linux. The Service Fabric cluster can run on-premise, on Azure, or on any cloud (e.g., AWS). You need to have at least five nodes to run a Service Fabric cluster for production workloads.

Service Fabric has the following system services to provide the platform capabilities.

Naming Service

A naming service resolves the service name to a location. Since applications in a cluster can move from one node to another, a naming service provides the actual port and IP address of the machine where the service is running.

Image Store Service

When performing a deployment, the application packages are uploaded to an image store, and then an application type is registered for the uploaded application package.

Failover Manager Service

As the name suggests, the failover manager service is responsible for the high availability and consistency of services. It orchestrates application and cluster upgrades.

Repair Manager Service

The repair manager service is an optional service to perform repair actions on silver and gold durability Azure Service Fabric clusters.

Cluster on Azure

A Service Fabric cluster on Azure can be created via the Azure portal or by using a resource template. Using the Azure portal user interface, a Service Fabric cluster can be easily created. Since the cluster and its components are like any other resource manager resource, you can easily track access, cost, and billing.

There are two major advantages of hosting a Service Fabric cluster on Azure.

- It comes with autoscaling functionality.

- It supports installation of Service Fabric clusters on Linux machines.

Standalone Cluster or Any Cloud Provider

Deployment on-premise or on any cloud provider is very similar. Service Fabric clusters can be created using the Windows Server 2012 R2 and Windows Server 2016 operating systems. Standalone clusters are useful in scenarios where you can't have your applications hosted on the cloud due to regulatory or compliance constraints.

Note   At the time of writing this book, Linux is not supported for standalone clusters. Linux can be used on one box for development purposes.

Develop and Deploy Applications on Service Fabric

So far in this chapter, we have discussed Service Fabric and its programming models, and learned that it can be installed on the cloud or on-premise. Now let's create a few samples to better understand how you actually develop and deploy applications on Service Fabric. We will cover two samples here.

- **Scenario 1**. Demonstrate developing an ASP.NET Core stateless web app communicating with an ASP.NET Core stateful API.

- **Scenario 2**. Demonstrate developing a Java Spring Boot application using Visual Studio Code and deploy it on Service Fabric as a guest executable or as a container.

Develop an ASP.NET Core Stateless Web App

We will develop a simple ASP.NET MVC–based application to manage employees. The ASP.NET MVC front end interacts with the ASP.NET API to perform CRUD operations. Internally, the Web API uses reliable collections to store employee data.

Setting up the Development Environment

Let's get started.

1. Install Visual Studio 2017.

2. Install the Microsoft Azure Service Fabric SDK.

3. Make sure that the Service Fabric local cluster is in a running state. Ensure this by browsing `http://localhost:19080/Explorer/index.html#/` or by right-clicking the Service Fabric icon in the system tray, as shown in Figure 3-7.

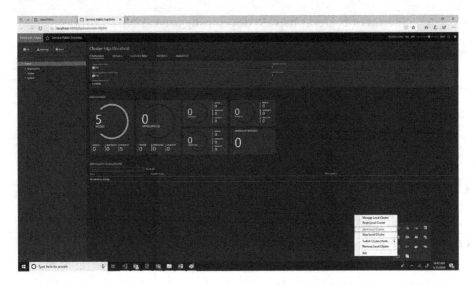

Figure 3-7. *Service Fabric status*

Create a ASP.NET Core Web API Using Reliable Collections

Here are the steps.

1. Launch Visual Studio 2017 as an administrator.

2. Create a project by selecting File ➤ New ➤ Project.

3. In the New Project dialog, choose Cloud ➤ Service Fabric Application.

4. Name the Service Fabric application **Employee** (as depicted in Figure 3-8) and click OK.

Figure 3-8. *New Service Fabric application*

5. Choose Stateful ASP.NET Core, as depicted in Figure 3-9.

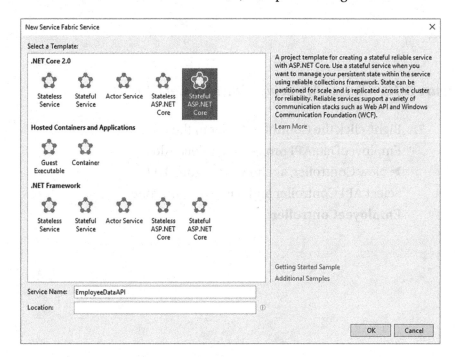

Figure 3-9. *New Stateful ASP.NET Core API*

6. You see a screen that looks like Figure 3-10.
    Click OK.

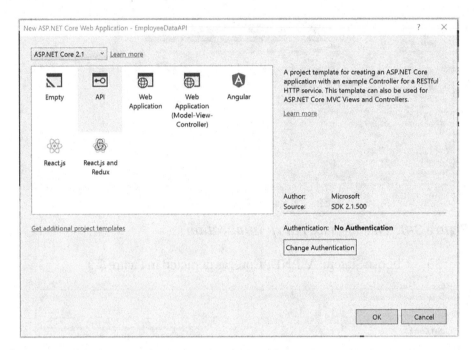

Figure 3-10.   *Choose API using (ASP.NET Core 2.1)*

7. Right-click the Controller folder in the
    EmployeeDataAPI project and select Add
    ➤ New Controller, as shown in Figure 3-11.
    Select API Controller and name the controller
    EmployeeController.

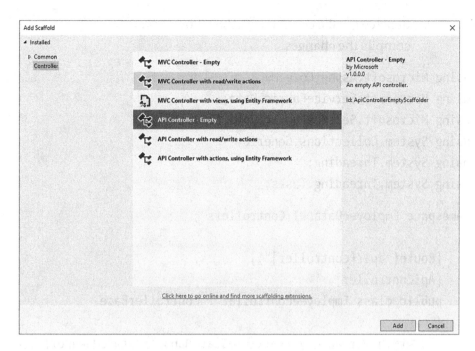

Figure 3-11. *New API controller*

8. Make sure that the NuGet packages shown in
   Figure 3-12 are installed.

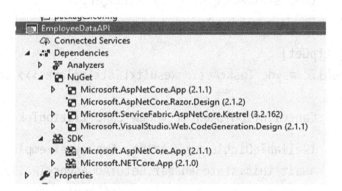

Figure 3-12. *NuGet Packages*

9. Replace the file content with the following and compile the changes.

```
using Microsoft.AspNetCore.Mvc;
using Microsoft.ServiceFabric.Data;
using Microsoft.ServiceFabric.Data.Collections;
using System.Collections.Generic;
using System.Threading;
using System.Threading.Tasks;

namespace EmployeeDataAPI.Controllers
{
    [Route("api/[controller]")]
    [ApiController]
    public class EmployeeController : ControllerBase
    {
        private readonly IReliableStateManager stateManager;

        public EmployeeController(IReliableStateManager
        stateManager)
        {
            this.stateManager = stateManager;
        }

        [HttpGet]
        public async Task<ActionResult<List<Employee>>> GetAll()
        {
            CancellationToken ct = new CancellationToken();

            IReliableDictionary<string, Employee> employees =
            await this.stateManager.GetOrAddAsync<IReliable
            Dictionary<string, Employee>>("employees");
```

```
List<Employee> employeesList = new
List<Employee>();

using (ITransaction tx = this.stateManager.
CreateTransaction())
{
    Microsoft.ServiceFabric.Data.IAsyncEnumerable
    <KeyValuePair<string, Employee>> list = await
    employees.CreateEnumerableAsync(tx);

    Microsoft.ServiceFabric.Data.IAsyncEnumerator
    <KeyValuePair<string, Employee>> enumerator =
    list.GetAsyncEnumerator();

    while (await enumerator.MoveNextAsync(ct))
    {
        employeesList.Add(enumerator.Current.Value);
    }
}

    return new ObjectResult(employeesList);
}

[HttpGet("{id}")]
public async Task<ActionResult<Employee>>
GetEmployee(string id)
{
    IReliableDictionary<string, Employee> employees =
    await this.stateManager.GetOrAddAsync<IReliable
    Dictionary<string, Employee>>("employees");

    Employee employee = null;
```

```
        using (ITransaction tx = this.stateManager.
        CreateTransaction())
        {
            ConditionalValue<Employee> currentEmployee =
            await employees.TryGetValueAsync(tx, id);

            if (currentEmployee.HasValue)
            {
                employee = currentEmployee.Value;
            }
        }

        return new OkObjectResult(employee);

    }

    [HttpPost]
    public async Task<ActionResult> Post(Employee employee)
    {
        IReliableDictionary<string, Employee> employees =
        await this.stateManager.GetOrAddAsync<IReliable
        Dictionary<string,Employee>>("employees");

        using (ITransaction tx = this.stateManager.
        CreateTransaction())
        {
            ConditionalValue<Employee> currentEmployee =
            await employees.TryGetValueAsync(tx, employee.
            Id.ToString());

            if (currentEmployee.HasValue)
            {
                await employees.SetAsync(tx, employee.
                Id.ToString(), employee);
```

```csharp
        }else
        {
            await employees.AddAsync(tx, employee.
            Id.ToString(), employee);
        }

        await tx.CommitAsync();
    }

    return new OkResult();
}

[HttpDelete("{id}")]
public async Task<ActionResult> Delete(string id)
{
    IReliableDictionary<string, Employee> employees =
    await this.stateManager.GetOrAddAsync<IReliable
    Dictionary<string, Employee>>("employees");

    using (ITransaction tx = this.stateManager.
    CreateTransaction())
    {
        if (await employees.ContainsKeyAsync(tx, id))
        {
            await employees.TryRemoveAsync(tx, id);

            await tx.CommitAsync();

            return new OkResult();
        }
        else
        {
            return new NotFoundResult();
        }
```

```
            }
        }

    }

    public class Employee
    {
        public string Name { get; set; }

        public string Mobile { get; set; }

        public long Id { get; set; }

        public string Designation { get; set; }
    }
}
```

Service Fabric provides a reliable collection in the form of reliable queues and a reliable dictionary. By using these classes, Service Fabric makes sure that the state is partitioned, replicated, and transacted within a partition.

Also, all the operations in a reliable dictionary object require an ITransaction object. By default, the Visual Studio template uses range partitioning; you can see the details in ApplicationManifest.xml, which resides in the ApplicationPackageRoot folder of the Employee project.

By default, the partition count is set to 1. The replica count is set to 3, which means a copy of the service code and data will be deployed on three nodes. Only one copy is active, called the *primary*, and the other two are inactive and used only in case of failure.

Here is a snippet from ApplicationManifest.xml for your reference.

```
<Parameter Name="EmployeeDataAPI_PartitionCount"
DefaultValue="1" />
<Parameter Name="EmployeeDataAPI_TargetReplicaSetSize"
DefaultValue="3" />
<Service Name="EmployeeDataAPI" ServicePackageActivationMode=
"ExclusiveProcess">
      <StatefulService ServiceTypeName="EmployeeDataAPIType"
      TargetReplicaSetSize="[EmployeeDataAPI_
      TargetReplicaSetSize]" MinReplicaSetSize="[EmployeeData
      API_MinReplicaSetSize]">
        <UniformInt64Partition PartitionCount="[EmployeeData
        API_PartitionCount]" LowKey="-9223372036854775808"
        HighKey="9223372036854775807" />
      </StatefulService>
   </Service>
```

Create an ASP.NET Web App Communicating with a Web API Using Proxy

Now we will create an ASP.NET-based web app that captures employee information and invokes EmployeeDataAPI using Service Fabric's reverse proxy.

1. Right-click the Services node under the Employee project and select New Service Fabric Service, as shown in Figure 3-13.

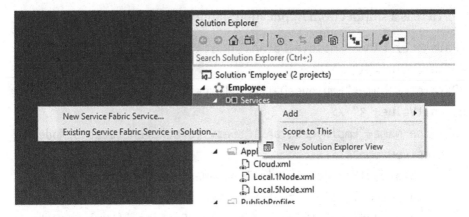

Figure 3-13. *New Service Fabric service*

2. In New Service Fabric Service, choose Stateless
    ASP.NET Core and name it **EmployeeWeb**.

3. In New ASP.NET Core Web Application, choose Web
    Application (Model-View-Controller) and click OK.

4. Create a class called EmployeeModel in the Models
    folder and add the following code.

```
using System;
using System.Collections.Generic;
using System.ComponentModel.DataAnnotations;
using System.Linq;
using System.Threading.Tasks;

namespace EmployeeWeb.Models
{
    public class EmployeeViewModel
    {
        public Employee Employee { get; set; }

        public List<Employee> EmployeeList { get; set; }
    }
```

```
public class Employee
{
    [Required]
    public string Name { get; set; }

    [Required]
    public string Mobile { get; set; }

    public long Id { get; set; }

    [Required]
    public string Designation { get; set; }
}
}
```

5. Replace the Index.cshtml content, which resides in
    the Views/Home folder, with the following content.

```
@model EmployeeViewModel

@{
    ViewData["Title"] = "Sample Employee Interface";
}

<div>
    <div class="container-fluid">
        <div class="row">
            <div class="col-xs-8 col-xs-offset-2 text-center">
                <h2>Sample Employee Interface</h2>
            </div>
        </div>

        <div class="row">
            <div class="col-xs-8 col-xs-offset-2">
                <form asp-action="Create" class="col-xs-12
                center-block">
```

```html
<div class="col-xs-6 form-group">
    <div class="form-group">
        <label asp-for="Employee.Name"
        class="control-label"></label>
        <input asp-for="Employee.Name"
        maxlength="100" class="form-
        control" />
        <span asp-validation-for="Employee.
        Name" class="text-danger"></span>
    </div>
    <div class="form-group">
        <label asp-for="Employee.
        Designation" class="control-label">
        </label>
        <input asp-for="Employee.
        Designation" maxlength="100"
        class="form-control" />
        <span asp-validation-for="Employee.
        Designation" class="text-danger">
        </span>
    </div>
    <div class="form-group">
        <label asp-for="Employee.Mobile"
        class="control-label"></label>
        <input asp-for="Employee.Mobile"
        maxlength="10" class=
        "form-control" />
        <span asp-validation-for="Employee.
        Mobile" class="text-danger"></span>
    </div>
```

```
              <div class="form-group">
                  <input type="submit" value="Create"
                  class="btn btn-default" />
              </div>
          </div>
      </form>
  </div>
</div>
<hr />
<div class="row">
    <div class="col-xs-8 col-xs-offset-2">
        @foreach (var item in Model.EmployeeList)
        {
            <div class="row">
                <div class="col-xs-2">
                    @Html.DisplayFor(modelItem =>
                    item.Id)
                </div>
                <div class="col-xs-2">
                    @Html.DisplayFor(modelItem =>
                    item.Name)
                </div>
                <div class="col-xs-2">
                    @Html.DisplayFor(modelItem =>
                    item.Designation)
                </div>
                <div class="col-xs-2">
                    @Html.DisplayFor(modelItem =>
                    item.Mobile)
                </div>
```

```
                        <div class="col-xs-2">
                            <a asp-action="Delete" asp-route-
                            id="@item.Id">Delete</a>
                        </div>
                    </div>
                }
            </div>
        </div>
    </div>
</div>
```

6. Replace the HomeController content with the following content.

```
using System;
using System.Collections.Generic;
using System.Diagnostics;
using System.Linq;
using System.Threading.Tasks;
using Microsoft.AspNetCore.Mvc;
using EmployeeWeb.Models;
using System.Net.Http;
using System.Fabric;
using EmployeeWeb.Proxy;

namespace EmployeeWeb.Controllers
{
    /// <summary>
    /// Employee Management Controller
    /// </summary>
    public class HomeController : Controller
    {
        private ServiceContext _serviceContext = null;
```

```csharp
private readonly HttpClient _httpClient;
private readonly FabricClient _fabricClient;
private static long EmployeeId = 0;

public HomeController(HttpClient httpClient,
StatelessServiceContext context, FabricClient
fabricClient)
{
    this._fabricClient = fabricClient;
    this._httpClient = httpClient;
    this._serviceContext = context;
}

/// <summary>
/// Loads the employee list
/// </summary>
/// <returns></returns>
public async Task<IActionResult> Index()
{
    EmployeeDataAPIProxy employeeProxy = new
    EmployeeDataAPIProxy(this._serviceContext,
    this._httpClient, this._fabricClient);

    List<Employee> employees = await employeeProxy.
    GetEmployees();

    EmployeeViewModel viewModel = new
    EmployeeViewModel();

    viewModel.EmployeeList = employees;

    return View(viewModel);
}
```

```
/// <summary>
/// Responsible for creating Employee
/// </summary>
/// <param name="employeeViewModel"></param>
/// <returns></returns>
[HttpPost]
[ValidateAntiForgeryToken]
public async Task<IActionResult>
Create(EmployeeViewModel employeeViewModel)
{
    EmployeeDataAPIProxy employeeProxy = new
    EmployeeDataAPIProxy(this._serviceContext,
    this._httpClient, this._fabricClient); ;

    if (ModelState.IsValid)
    {
        //Not for production at all.
        EmployeeId = EmployeeId + 1;

        employeeViewModel.Employee.Id = EmployeeId;

        await employeeProxy.CreateEmployee(employeeView
        Model.Employee);

    }

    List<Employee> employees = await employeeProxy.
    GetEmployees();

    employeeViewModel.EmployeeList = employees;

    return View("Index",employeeViewModel);
}
/// <summary>
/// Delete
```

```
/// </summary>
/// <param name="id"></param>
/// <returns></returns>
[HttpGet, ActionName("Delete")]
public async Task<IActionResult> Delete(long? id)
{
    EmployeeViewModel viewModel = new
    EmployeeViewModel();

    EmployeeDataAPIProxy employeeProxy = new
    EmployeeDataAPIProxy(this._serviceContext,
    this._httpClient, this._fabricClient);

    if (id != null)
    {
        await employeeProxy.DeleteEmployee(id.Value);
    }

    List<Employee> employees = await employeeProxy.
    GetEmployees();

    viewModel.EmployeeList = employees;

    return View("Index", viewModel);
}

}
}
```

7. Create a new folder called Proxy in the
   EmployeeWeb project. Add a new file called
   EmployeeDataAPIProxy.cs.

This code is responsible for invoking the EmployeeDataAPI. Since EmployeeDataAPI is a stateful service with one partition and three replicas, we use reverse proxy to find out the active replica. The reverse proxy is configured by default to use port 19081. You can confirm the port by looking at the HttpApplicationGatewayEndpoint port in your local cluster manifest.

Also, since it's a sample application, we used a running sequence number as a partition key while invoking the stateful service. Please add the following code to EmployeeDataAPIProxy.cs.

```
using EmployeeWeb.Models;
using Newtonsoft.Json;
using System;
using System.Collections.Generic;
using System.Fabric;
using System.Fabric.Query;
using System.Linq;
using System.Net.Http;
using System.Net.Http.Headers;
using System.Threading.Tasks;

namespace EmployeeWeb.Proxy
{
    /// <summary>
    /// Proxy class to handle the complexity of dealing with
    reliable service
    /// </summary>
    public class EmployeeDataAPIProxy
    {
        private ServiceContext _context = null;
        private readonly HttpClient _httpClient;
        private readonly FabricClient _fabricClient;
```

```csharp
public EmployeeDataAPIProxy(ServiceContext context,
HttpClient httpClient,FabricClient fabricClient)
{
    _context = context;

    _httpClient = httpClient;

    _fabricClient = fabricClient;
}

/// <summary>
/// Returns the list of employees from all the
partitions. In our sample we have only 1 partition
/// Also, we are making use of proxy to determine the
right partition to connect to.
/// Please refer this link for more details. https://
docs.microsoft.com/en-us/azure/service-fabric/service-
fabric-reverseproxy
/// </summary>
/// <returns></returns>
public async Task<List<Employee>> GetEmployees()
{
    Uri serviceName = EmployeeWeb.
    GetEmployeeDataServiceName(_context);

    Uri proxyAddress = this.
    GetProxyAddress(serviceName);

    ServicePartitionList partitions = await
    _fabricClient.QueryManager.GetPartitionListAsync
    (serviceName);

    List<Employee> employees = new List<Employee>();
```

```
        foreach (Partition partition in partitions)
        {
            string proxyUrl =
                $"{proxyAddress}/api/Employee?Partition
                Key={((Int64RangePartitionInformation)
                partition.PartitionInformation).LowKey}&
                PartitionKind=Int64Range";

            using (HttpResponseMessage response = await
            _httpClient.GetAsync(proxyUrl))
            {
                if (response.StatusCode != System.Net.
                HttpStatusCode.OK)
                {
                    continue;
                }

                employees.AddRange(JsonConvert.Deserialize
                Object<List<Employee>>(await response.
                Content.ReadAsStringAsync()));
            }
        }

        return employees;
    }

    /// <summary>
    /// Creates an Employee
    /// </summary>
    /// <param name="employee"></param>
    /// <returns></returns>
```

```csharp
public async Task CreateEmployee(Employee employee)
{
    Uri serviceName = EmployeeWeb.
    GetEmployeeDataServiceName(_context);

    Uri proxyAddress = this.
    GetProxyAddress(serviceName);

    long partitionKey = employee.Id;

    string proxyUrl = $"{proxyAddress}/api/Employee?
    PartitionKey={partitionKey}&PartitionKind=Int64Range";

    await this._httpClient.PostAsJsonAsync<Employee>
    (proxyUrl, employee);
}

/// <summary>
/// Deletes an Employee
/// </summary>
/// <param name="employee"></param>
/// <returns></returns>
public async Task DeleteEmployee(long employeeId)
{
    Uri serviceName = EmployeeWeb.
    GetEmployeeDataServiceName(_context);

    Uri proxyAddress = this.GetProxyAddress(serviceName);

    long partitionKey = employeeId;

    string proxyUrl = $"{proxyAddress}/api/Employee/
    {employeeId}?PartitionKey={partitionKey}&Partition
    Kind=Int64Range";

    await this._httpClient.DeleteAsync(proxyUrl);
}
```

```
/// <summary>
/// Constructs a reverse proxy URL for a given service.
/// To find the reverse proxy port used in
your local development cluster, view the
HttpApplicationGatewayEndpoint element in the local
Service Fabric cluster manifest:
/// Open a browser window and navigate to http://
localhost:19080 to open the Service Fabric Explorer
tool.
/// Select Cluster -> Manifest.
/// Make a note of the HttpApplicationGatewayEndpoint
element port.By default this should be 19081. If
it is not 19081, you will need to change the port
in the GetProxyAddress method of the following
VotesController.cs code.
/// </summary>
/// <param name="serviceName"></param>
/// <returns></returns>
private Uri GetProxyAddress(Uri serviceName)
{
    return new Uri($"http://localhost:19081
{serviceName.AbsolutePath}");
}
}
}
```

8. Replace the CreateServiceInstanceListeners()
   function with the following code in EmployeeWeb.cs.

```csharp
/// <summary>
    /// Optional override to create listeners (like tcp,
    http) for this service instance.
    /// </summary>
    /// <returns>The collection of listeners.</returns>
    protected override IEnumerable<ServiceInstanceListener>
    CreateServiceInstanceListeners()
    {
        return new ServiceInstanceListener[]
        {
            new ServiceInstanceListener(serviceContext =>
                new KestrelCommunicationListener(service
                Context, "ServiceEndpoint", (url, listener) =>
                {
                    ServiceEventSource.Current.
                    ServiceMessage(serviceContext,
                    $"Starting Kestrel on {url}");

                    return new WebHostBuilder()
                            .UseKestrel()
                            .ConfigureServices(
                                services => services
                                    .AddSingleton
                                    <HttpClient>(new
                                    HttpClient())
                                    // Add this line to
                                    default template code
                                    .AddSingleton
                                    <FabricClient>(new
                                    FabricClient())
                                    //Add this line to
                                    default template code
```

```
                                        .AddSingleton<State
                                        lessServiceContext>
                                        (serviceContext))
                                    .UseContentRoot(Directory.
                                    GetCurrentDirectory())
                                    .UseStartup<Startup>()
                                    .UseServiceFabricIntegration
                                    (listener, ServiceFabric
                                    IntegrationOptions.None)
                                    .UseUrls(url)
                                    .Build();
                }))
        };
    }
```

9. Add the following private function to EmployeeWeb.cs.

```
internal static Uri GetEmployeeDataServiceName(ServiceContext
context)
        {
            return new Uri($"{context.CodePackageActivation
            Context.ApplicationName}/EmployeeDataAPI");
        }
```

10. Make sure that the following namespaces are
     present in EmployeeWeb.cs at the top of the class
     file.

```
using System;
using System.Collections.Generic;
using System.Fabric;
using System.IO;
using Microsoft.AspNetCore.Hosting;
```

```
using Microsoft.Extensions.DependencyInjection;
using Microsoft.ServiceFabric.Services.Communication.
AspNetCore;
using Microsoft.ServiceFabric.Services.Communication.Runtime;
using Microsoft.ServiceFabric.Services.Runtime;
using System.Net.Http;
```

Debugging the Application

By executing all the steps in the previous section, your development is complete. The following steps debug the application to create an employee record in the Service Fabric reliable collection that utilizes the developed web interface and data API.

1. Right-click the Employee project and set the Application URL to "http://localhost:19080/ Explorer". By default, Service Fabric Explorer runs on 19080. This ensures the successful deployment of the service to a local cluster. It launches Service Fabric Explorer.

2. Make sure that the Employee project is set at start up.

3. Click F5. This deploys your Service Fabric application to the local development cluster.

4. In Service Fabric Explorer, click Application. Click fabric://Employee, fabric://Employee/ EmployeWeb, Partition ID, and Node ID. Copy the value of the endpoint. (In our case, EmployeeWeb is hosted at http://localhost:8780.)

5. You can also get the EmployeeWeb port number from ServiceManifest.xml.

```
<Endpoints>
<!-- This endpoint is used by the communication listener to
obtain the port on which to listen. Please note that if your
service is partitioned, this port is shared with replicas of
different partitions that are placed in your code. -->
<Endpoint Protocol="http" Name="ServiceEndpoint" Type="Input"
Port="8780" />
</Endpoints>
```

6. Browse the `http://localhost:8780/` URL to view the web interface. Enter employee information and click Create to create an employee record. The actual data is saved by the Employee Data API in the Service Fabric's reliable collection, instead of an external database like Azure SQL.

Develop a Spring Boot Application

In the previous example, we developed ASP.NET-based reliable services and deployed it to a local Service Fabric cluster. In this example, we showcase that it is possible to host a non-Microsoft stack application to a Service Fabric cluster. We will not create a reliable service; instead, we will host a Java Spring Boot–based API as a guest executable and as a container. We will use VS Code to develop a simple Spring Boot application.

Setting up the Development Environment

Let's set up the development environment.

1. Install Visual Studio 2017.

2. Install the Microsoft Azure Service Fabric SDK.

3. Install Visual Studio Code.

    a. Install Spring Boot Extensions Pack.

    b. Install Java Extensions Pack.

    c. Install Maven for Java.

4. Make sure that the Service Fabric Local cluster is in a running state.

5. Install Docker Desktop.

6. Access the Azure container registry.

Develop a Spring Boot API

Now it's time to get started on the application.

1. Launch Visual Studio Code as an administrator.

2. Press Ctrl+Shift+P to open the command palette.

3. Enter **spring** in the command palette, as shown in Figure 3-14, and choose Spring Initializr: Generate a Maven Project.

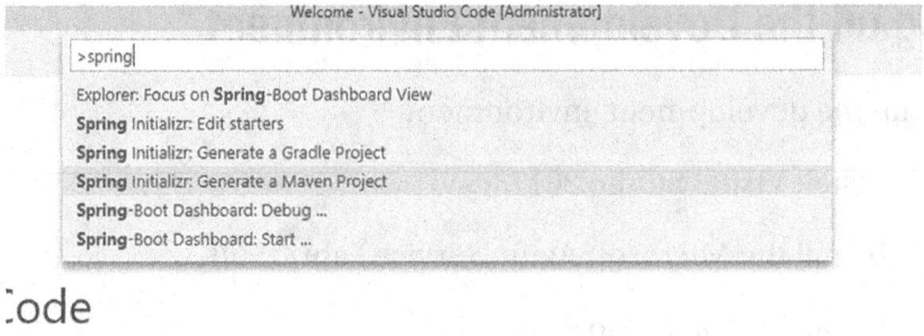

Welcome - Visual Studio Code [Administrator]

> spring|

Explorer: Focus on **Spring**-Boot Dashboard View
Spring Initializr: Edit starters
Spring Initializr: Generate a Gradle Project
Spring Initializr: Generate a Maven Project
Spring-Boot Dashboard: Debug ...
Spring-Boot Dashboard: Start ...

:ode

Figure 3-14. *Visual Studio Code command palette*

4. Choose Java for Specify Project Language.

5. Enter **com.microservices** in the input group ID for your project.

6. Enter **employeespringservice** in the input artifact ID for your project.

7. Choose the latest Spring boot version (at the time of writing it was 2.1.2).

8. Choose the following dependencies.

   a. DevTools

   b. Lombok

   c. Web

   d. Actuator

9. Choose the path where you want to save the solution.

10. Right-click the EMPLOYEESPRINGSERVICE folder under src ➤ main ➤ java ➤ com ➤ microservices, as shown in Figure 3-15, and click Add File.

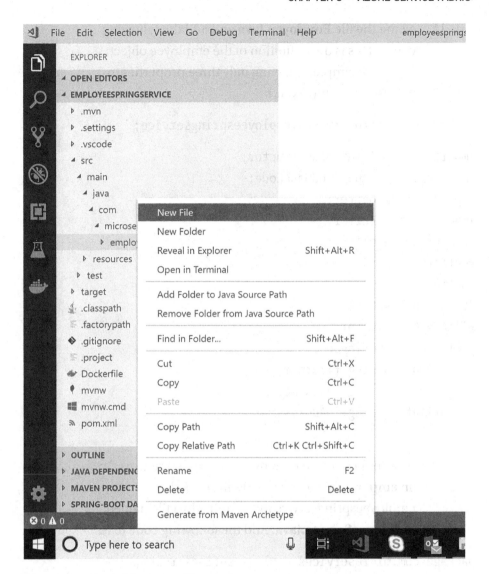

Figure 3-15. *Add a new file*

11. Name the file **Employee.Java** and add the following
    code. (This is the definition of the employee object;
    we kept it simple by having only three properties to
    represent an employee.)

```
package com.microservices.employeespringservice;

import lombok.AllArgsConstructor;
import lombok.EqualsAndHashCode;
import lombok.Getter;
import lombok.Setter;
/*** Employee ***/
@Getter
@Setter
@EqualsAndHashCode
@AllArgsConstructor
public class Employee {
    private String firstName;
    private String lastName;
    private String ipAddress;
}
```

12. Now let's create an employee service that returns
    an employee's information. Right-click the
    employeespringservice folder and add a file named
    EmployeeService.java. Add the following code to it.

```
package com.microservices.employeespringservice;

import java.net.InetAddress;
import java.net.UnknownHostException;

import org.springframework.stereotype.Service;
```

```
/**
 * EmployeeService
 */
@Service
public class EmployeeService {

    public Employee GetEmployee(String firstName, String
    lastName){

        String ipAddress;

        try {

            ipAddress = InetAddress.getLocalHost().
            getHostAddress().toString();

        } catch (UnknownHostException e) {

            ipAddress = e.getMessage();
        }

        Employee employee = new Employee(firstName,
        lastName,ipAddress);

        return employee;
    }
}
```

13. Now let's create an employee controller that invokes
    the employee service to return the details of an
    employee. Right-click the employeespringservice
    folder and add a file named **EmployeeController.
    java**. Add the following code to it.

```java
package com.microservices.employeespringservice;

import org.springframework.beans.factory.annotation.Autowired;
import org.springframework.stereotype.Controller;
import org.springframework.web.bind.annotation.GetMapping;
import org.springframework.web.bind.annotation.ResponseBody;

/**
 * EmployeeController
 */
@Controller
public class EmployeeController {

    @Autowired
    private EmployeeService employeeService;

    @GetMapping("/")
    @ResponseBody
    public Employee getEmployee(){
        return employeeService.GetEmployee("Spring","Boot");
    }
}
```

Now you are ready for a simple REST-based service that returns employee information. Visual Studio Code has some cool features to run Spring Boot applications, and we are going to use the same.

1. Open DemoApplication.java and once you open the file, you see the option to run or debug the application, as shown in Figure 3-16. Please note that this may take time as Visual Studio automatically downloads the dependencies.

```
⚲ DemoApplication.java ✕
 1    package com.microservices.employeespringservice;
 2
 3    import org.springframework.boot.SpringApplication;
 4    import org.springframework.boot.autoconfigure.SpringBootApplication;
 5
 6    @SpringBootApplication
 7    public class DemoApplication {
 8
      ┌─────────┐
      │Run│Debug│
      └─────────┘
 9        public static void main(String[] args) {
10            SpringApplication.run(DemoApplication.class, args);
11        }
12
13    }
14
15    |
```

Figure 3-16. *Debug Application*

2. Click Run and open the Controller class. You see
    the URL where your controller service is hosted, as
    shown in Figure 3-17.

```
nal  Help                     EmployeeController.java - employeespringservice - Visual Studio Code
   DemoApplication.java      ⠿   ❚❚   ↷   ↓   ↑   ↻   ■
     1   package com.microservices.employeespringservice;
     2
     3   import org.springframework.beans.factory.annotation.Autowired;
     4   import org.springframework.stereotype.Controller;
     5   import org.springframework.web.bind.annotation.GetMapping;
     6   import org.springframework.web.bind.annotation.ResponseBody;
     7
     8   /**
     9    * EmployeeController
    10    */
          ← EmployeeService
    11   @Controller
    12   public class EmployeeController {
    13
              ← EmployeeService
    14       @Autowired
    15       private EmployeeService employeeService;
    16
          http://127.0.0.1:8080/
    17       @GetMapping("/")
    18       @ResponseBody
    19       public Employee getEmployee(){
    20           return employeeService.GetEmployee("Spring","Boot");
    21       }
    22   }
```

Figure 3-17. *Application URL*

3. Click the URL and you see the output shown in
    Figure 3-18 in your default browser.

{"firstName":"Spring","lastName":"Boot","ipAddress":"10.0.75.1"}

Figure 3-18. *Application output*

4. Right-click employeespringservice under Maven
   Projects. Click package, as shown in Figure 3-19.
   This generates the JAR file.

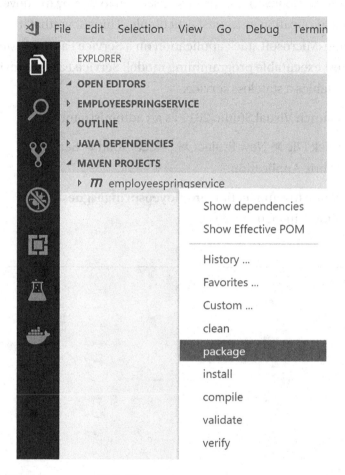

Figure 3-19. *Generate JAR file*

Now you have a simple Spring Boot-based REST API, and you have
generated a JAR file. We will now deploy it to Service Fabric as a guest
executable. To deploy, we will use Visual Studio 2017.

Deploy a Spring Boot Service as a Guest Executable

After executing all the steps in the previous section, your development is complete. Please follow the steps in this section to deploy the developed Spring Boot application as a guest executable. This shows that it is possible to host a non-Microsoft stack application on a Service Fabric cluster by using a guest executable programming model. Service Fabric considers guest executables a stateless service.

1. Launch Visual Studio 2017 as an administrator.

1. Click File ➤ New Project ➤ Select Cloud ➤ Service Fabric Application.

2. Name the application **employeespringasguest**, as shown in Figure 3-20.

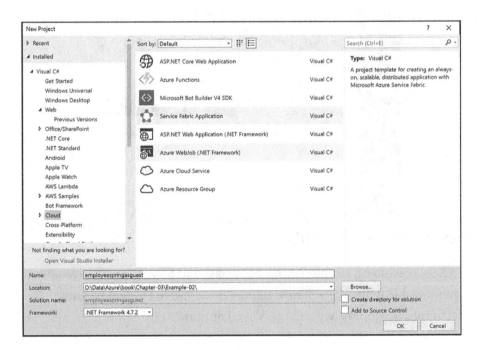

Figure 3-20. *Create Service Fabric application*

94

3. In New Service Fabric Service, select the following
    (as shown in Figure 3-21).

    a. Service Name: employeeguestservice

    b. Code Package Folder: Point to the target folder in which
        Visual Studio Code generated the JAR file for the Spring Boot
        service.

    c. Code Package Behavior: Copy folder contents to folder

    d. Working Folder: CodeBase

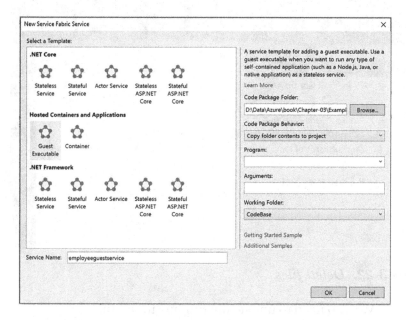

Figure 3-21. *New Service Fabric Service*

4. Delete the selected files shown in Figure 3-22 from
    the Code folder.

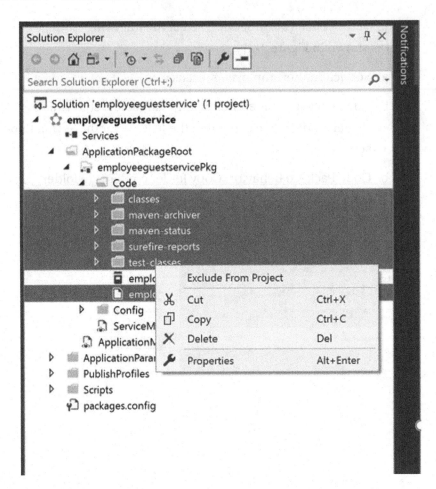

Figure 3-22. *Delete files*

5. We also need to upload the runtime to run the
   JAR. Generally, it resides in the JDK installation
   folder (C:\ java-1.8.0-openjdk-1.8.0.191-1.b12.
   redhat.windows.x86_64). Paste it in the Code folder,
   as shown in Figure 3-23.

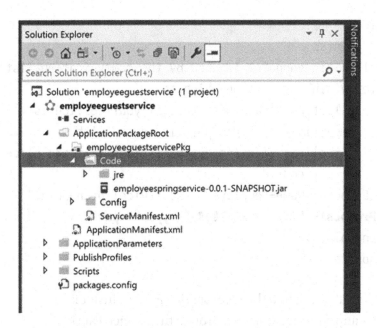

Figure 3-23. *Adding Java runtime*

6. Open ServiceManisfest.xml and set the following values.

```
<EntryPoint>
    <ExeHost>
        <Program>jre\bin\java.exe</Program>
        <Arguments>-jar ..\..\employeespringservice-0.0.1-
        SNAPSHOT.jar</Arguments>
        <WorkingFolder>CodeBase</WorkingFolder>
        <!-- Uncomment to log console output (both stdout and
        stderr) to one of the
            service's working directories. -->
        <!-- <ConsoleRedirection FileRetentionCount="5"
        FileMaxSizeInKb="2048"/> -->
    </ExeHost>
  </EntryPoint>
</CodePackage>
```

```
<Resources>
   <Endpoints>
      <!-- This endpoint is used by the communication listener
      to obtain the port on which to
            listen. Please note that if your service is
            partitioned, this port is shared with
            replicas of different partitions that are placed in
            your code. -->
      <Endpoint Name="employeeguestserviceTypeEndpoint"
      Protocol="http" Port="8080" Type="Input" />
   </Endpoints>
</Resources>
```

7. Make sure that the local Service Fabric cluster is up and running. Click F5. Browse the Service Fabric dashboard, as shown in Figure 3-24. The default URL is http://localhost:19080/Explorer/index. html. You see that your service is deployed.

Figure 3-24. *Service Fabric dashboard*

8. Browse http://localhost:8080 to access your
    service. In servicemanifest.xml, we specified the
    service port as 8080; you can browse the same on
    8080, as shown in Figure 3-25.

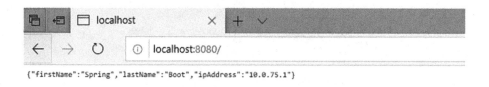

Figure 3-25. *Application output*

Deploy a Spring Boot Service as a Container

So far, we have deployed the service as a guest executable in Service Fabric.
Now we will follow the steps to deploy the Spring service as a container
in Service Fabric. This explains that in addition to creating stateful and
stateless services, Service Fabric also orchestrates containers like any other
orchestrator, even if the application wasn't developed on a Microsoft stack.

1. Open Visual Studio Code. Open the folder where
    employeespringservice exists. Open the Docker file.

2. Make sure that the name of the JAR file is correct.

3. Select **Switch to Windows container...** in Docker
    Desktop, as shown in Figure 3-26.

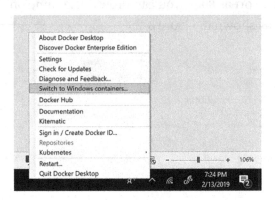

Figure 3-26. *Switch to Windows containers*

4. Create the Azure Container Registry resource in the
    Azure portal. Enable the admin user, as shown in
    Figure 3-27.

myservicefabric - Access keys
Container registry

- Search (Ctrl+/)

- Overview
- Activity log
- Access control (IAM)
- Tags
- Quick start
- Events

Settings

- Access keys
- Locks
- Automation script

Services

- Repositories
- Webhooks
- Replications

Policies

- Content trust (Preview)

Monitoring

- Metrics (Preview)

Support + troubleshooting

- New support request

Registry name

myservicefabric

Login server

myservicefabric.azurecr.io

Admin user ●
Enable   Disable

Username

myservicefabric

NAME	PASSWORD
password	
password2	

Figure 3-27. *Azure Key Vault*

5.   Open the command prompt in Administrative Mode and browse to the directory where the Docker file exists.

6.   Fire the following command, including the period at the end. (This may take time because it downloads the Window Server core image from the Docker hub, as shown in Figure 3-28.)

```
docker build -t employeespringservice/v1 .
```

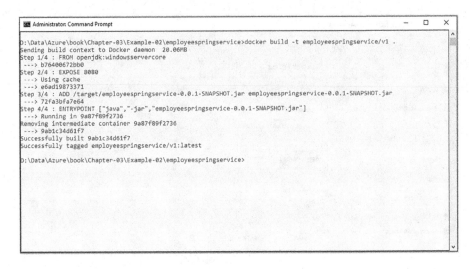

Figure 3-28. *Docker build output*

Now the container image is available locally. You have to push the image to Azure Container Registry.

1. Log in to Azure Container Registry using the admin username and password. Use the following command (also see Figure 3-29).

```
docker login youracr.azurecr.io -u yourusername -p yourpassword
```

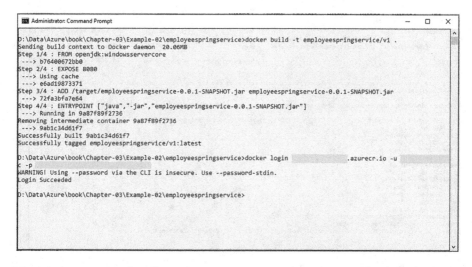

Figure 3-29. Docker login

2. Fire the following commands to upload the image to ACR (as shown in Figure 3-30).

docker tag employeespringservice/v1 youracr.azurecr.io/book/
employeespringservice/v1

docker push myservicefabric.azurecr.io/book/
employeespringservice/v1

```
Administrator: Command Prompt                                              —    □    ×
employeeservice/v1                    latest        154a12b32578   3 weeks ago    4.52 ^
GB
myservicefabric.azurecr.io/samples/employeeservice/v1   latest   154a12b32578   3 weeks ago   4.52
GB
openjdk                               windowsservercore   b76400672bb0   3 weeks ago    4.5G
B
D:\Data\Azure\book\Chapter-03\Example-02\employeespringservice>docker tag employeespringservice/v1 myservicefabric.azure
cr.io/book/employeespringservice/v1
D:\Data\Azure\book\Chapter-03\Example-02\employeespringservice>docker push              .azurecr.io/book/employeesprin
gservice/v1
The push refers to repository [              .azurecr.io/book/employeespringservice/v1]
3ba5f91bb947: Pushed
c309774bb359: Pushed
a07cca47fd47: Pushed
aff8ea0481d2: Pushed
664240f6bb32: Pushed
f9da363fd495: Pushed
6ed33c44b167: Pushed
98331e8a4501: Pushed
176c7b727a0b: Pushed
e308c396e652: Pushed
f0f9b327dc34: Pushed
6508d5ab77c5: Pushed
9aa5aa8919b7: Skipped foreign layer
c4d02418787d: Skipped foreign layer
latest: digest: sha256:d678e2e92abafdfe8b8172e08f8a5fe31bec1ca7151d2e2a5075adea8eaa54bf size: 3456

D:\Data\Azure\book\Chapter-03\Example-02\employeespringservice>
```

Figure 3-30. *Docker push*

3. Log in to the Azure portal and check if you can see
    your image in Repositories, as shown in Figure 3-31.

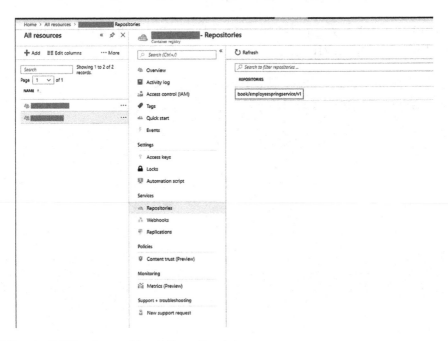

Figure 3-31. *Azure Container Registry*

Since the container image is ready and uploaded in Azure Container Registry, let's create a Service Fabric project to deploy the container to the local Service Fabric cluster.

1. Launch Visual Studio 2017 as an administrator.

2. Click File ➤ New Project ➤ Select Cloud ➤ Service Fabric Application.

3. Name the application **employeespringascontainer**, as shown in Figure 3-32.

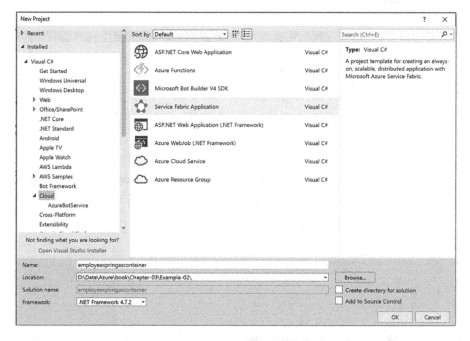

Figure 3-32. *Create Service Fabric application*

4. In New Service Fabric Service, select the following.

    a. Service Name: employeecontainerservice

    b. Image Name: youracr.azurecr.io/book/ employeespringservice/v1

105

   c. User Name: Your username in the Azure Container Registry

   d. Host Port: 8090

   e. Container Port: 8080

5. Once the solution is created, open the ApplicationManifest.xml. Specify the right password for the admin user (see Figure 3-33). (Since this is a sample, we kept the password unencrypted; for real-word applications you have to encrypt the password.)

```
ApplicationManifest.xml  ⇄ ×  ServiceManifest.xml
 1   <?xml version="1.0" encoding="utf-8"?>
 2   <ApplicationManifest ApplicationTypeName="employeespringascontainerType"
 3                        ApplicationTypeVersion="1.0.0"
 4                        xmlns="http://schemas.microsoft.com/2011/01/fabric"
 5                        xmlns:xsd="http://www.w3.org/2001/XMLSchema"
 6                        xmlns:xsi="http://www.w3.org/2001/XMLSchema-instance">
 7     <Parameters>
 8       <Parameter Name="employeecontainerservice_InstanceCount" DefaultValue="-1" />
 9     </Parameters>
10     <!-- Import the ServiceManifest from the ServicePackage. The ServiceManifestName and ServiceManifestVersion
11          should match the Name and Version attributes of the ServiceManifest element defined in the
12          ServiceManifest.xml file. -->
13     <ServiceManifestImport>
14       <ServiceManifestRef ServiceManifestName="employeecontainerservicePkg" ServiceManifestVersion="1.0.0" />
15       <ConfigOverrides />
16       <Policies>
17         <ContainerHostPolicies CodePackageRef="Code">
18           <!-- See https://aka.ms/I7z0p9 for how to encrypt your repository password -->
19           <RepositoryCredentials AccountName="myservicefabric" Password="            " PasswordEncrypted="false" />
20           <PortBinding ContainerPort="8080" EndpointRef="employeecontainerserviceTypeEndpoint" />
21         </ContainerHostPolicies>
22       </Policies>
23     </ServiceManifestImport>
24     <DefaultServices>
25       <!-- The section below creates instances of service types, when an instance of this
26            application type is created. You can also create one or more instances of service type using the
27            ServiceFabric PowerShell module.
28
29            The attribute ServiceTypeName below must match the name defined in the imported ServiceManifest.xml file. -->
30       <Service Name="employeecontainerservice" ServicePackageActivationMode="ExclusiveProcess">
31         <StatelessService ServiceTypeName="employeecontainerserviceType" InstanceCount="[employeecontainerservice_InstanceCount]">
32           <SingletonPartition />
33         </StatelessService>
34       </Service>
35     </DefaultServices>
36   </ApplicationManifest>
```

Figure 3-33. *Application manifest*

Now we are ready to build and deploy the container to the local Service Fabric cluster. Since we have given the user information to download the image from Azure Container Registry, Visual Studio downloads and deploys the container to the local Service Fabric cluster.

1. Right-click the Service Fabric project and publish, as
    shown in Figure 3-34.

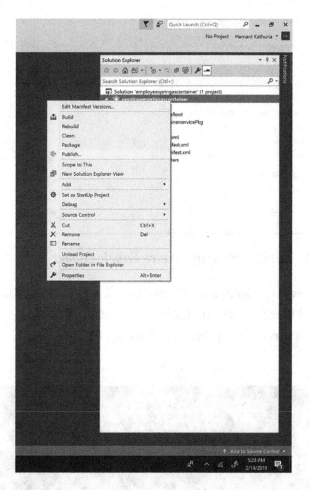

Figure 3-34. *Publish*

2. Select the local cluster profile to publish to the local
    Service Fabric cluster, as shown in Figure 3-35. To
    deploy to Azure, select the cloud profile. Make sure
    that the Service Fabric cluster is up and ready in
    your subscription.

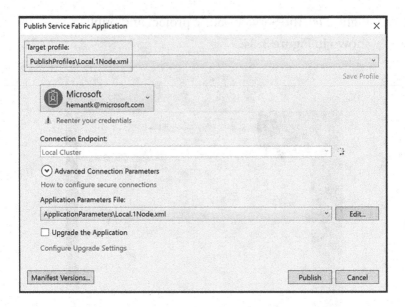

Figure 3-35.   *Publish to local Service Fabric cluster*

3.   Browse the Service Fabric dashboard. The default URL
     is `http://localhost:19080/Explorer/index.html`.
     Your service is deployed, as shown in Figure 3-36.

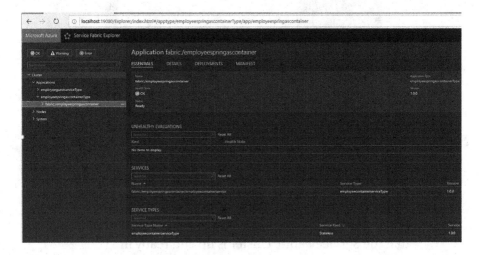

Figure 3-36.   *Service Fabric dashboard*

4. Browse to http://localhost:8090/ to access your
    service. You get the response shown in Figure 3-37,
    which is served from the container run by Service
    Fabric.

{"firstName":"Spring","lastName":"Boot","ipAddress":"172.30.144.129"}

Figure 3-37. *Application output*

Summary

Service Fabric is not just a Microsoft orchestrator to host containers for
microservices. In addition to host containers, it also allows to develop
reliable services by using the Service Fabric SDK. SDK allows you to
implement service communication, scale, and service discovery patterns
effectively. The SDK is available for .NET and Java developers, and the
cluster can be deployed on-premise or on the cloud.

Please note that it's not mandatory to use the Service Fabric SDK. You
can develop an application in any programming language, and you can
deploy it using guest executables and containers. You can deploy Service
Fabric on any cloud, such as Azure, AWS, or Google.

CHAPTER 4

Monitoring Azure Service Fabric Clusters

In the previous chapter, you learned the core concepts of Service Fabric, such as the application model, application scaling, supported programming models, and clusters. Service Fabric very effectively streamlines an application, cluster deployment, and the scaling of applications. In this chapter, you discover how you can effectively monitor a Service Fabric cluster and the applications deployed on it. We will create an ASP.NET Core-based application to demonstrate how easy it is to add application monitoring, and use Application Insights to troubleshoot issues in database calls and remote HTTP calls.

In an enterprise application, most performance issues happen in an inefficient database and remote HTTP calls. If an application does not have efficient monitoring and logging built in, it becomes very difficult to solve performance issues. Our sample app demonstrates how to monitor key performance issues.

Before we discuss monitoring, however, the following sections are brief descriptions of some of the technologies mentioned in this chapter.

© Harsh Chawla and Hemant Kathuria 2019
H. Chawla and H. Kathuria, *Building Microservices Applications on Microsoft Azure*,
https://doi.org/10.1007/978-1-4842-4828-7_4

Azure Application Insights

Application Insights is an extensible application performance management (APM) service for web developers on multiple platforms. It monitors live web applications and automatically detects performance anomalies. It also includes powerful analytics tools to help you diagnose issues and to understand what users do with your app.

Resource Manager Template

The Azure Resource Manager template allows you to deploy, monitor, and manage solution resources as a group on Azure.

We cover Service Fabric monitoring in the following three areas.

- Application monitoring

- Cluster monitoring

- Infrastructure monitoring

Application Monitoring

Application monitoring shows the usage of your application's features and components, which helps determine their impact on the users. Application monitoring also reports debug and exception logs, which are essential for diagnosing and resolving an issue once the application is deployed. It is the responsibility of developers to add appropriate monitoring.

You can use any popular instrumentation framework to add application monitoring, but some of the popular options are Application Insights SDK, Event Source, and ASP.NET Core Logging Framework.

Application Insights is recommended.

Adding Application Monitoring to a Stateless Service Using Application Insights

We will develop a simple ASP.NET MVC–based API to manage employees. In this example, we will store the employee data in an Azure SQL database instead of a reliable collection so that we can demonstrate how to monitor query information in Azure Application Insights. To demonstrate the monitoring of a REST API call, we are making a call to the Translator Text API in Azure to transliterate the first name of an employee in Hindi (Devanagari) script. You can replace the call with any other REST call, as the idea here is to demonstrate the monitoring of remote calls in Azure Application Insights. In your Azure subscription, you can create a Translator Text API using the Free tier to execute this sample.

Note   The Microsoft Translator API is an ISO and HIPAA-compliant neural machine translation (NMT) service that developers can easily integrate into their applications, websites, tools, or any solution requiring multilanguage support, such as company websites, e-commerce sites, customer support applications, messaging applications, internal communication, and more.

Setting up the Development Environment

Let's set up.

1. Install Visual Studio 2017.

2. Install the Microsoft Azure Service Fabric SDK.

3. Create the Translator Text API in your Azure subscription and make a note of the access key.

4. Create an empty Azure SQL Database and keep the connection string with SQL Authentication handy.

5. Make sure that the Service Fabric local cluster on Windows is in a running state.

6. Make sure that the Service Fabric Azure cluster on Windows is in a running state.

Create an ASP.NET Core Web API

Now let's start the API.

1. Launch Visual Studio 2017 as an administrator.

2. Create a project by selecting File ➤ New ➤ Project.

3. In the New Project dialog, choose Cloud ➤ Service Fabric Application.

4. Name the Service Fabric application **EmployeeApp** (as seen in Figure 4-1) and click OK.

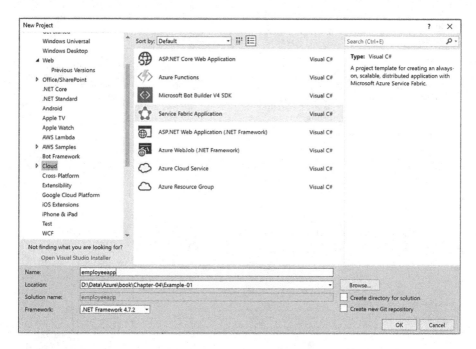

Figure 4-1. *Create Service Fabric application*

5. Name the stateless ASP.NET Core service **Employee.
    Stateless.Api** (as seen in Figure 4-2) and click OK.

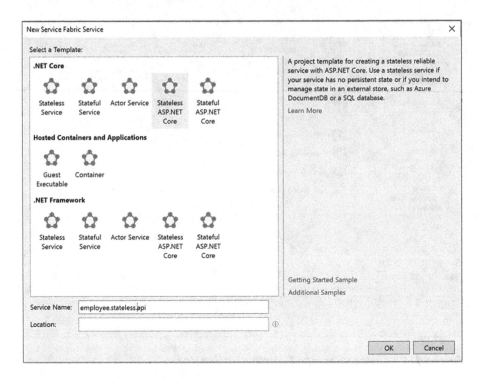

Figure 4-2. *Stateless ASP.NET Core*

6. Choose the API and click OK. Make sure that
    ASP.NET Core 2.2 is selected, as shown in Figure 4-3.

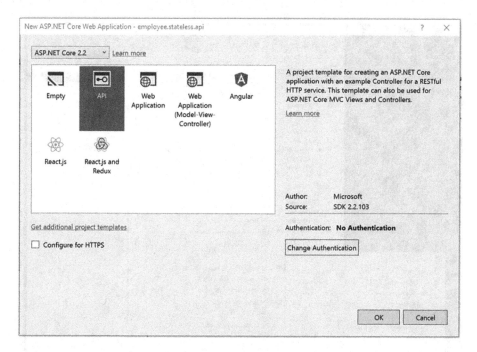

Figure 4-3. *API with ASP.NET Core 2.2*

7. Right-click the employee.stateless.api project
   and select Add ➤ Connected Service, as seen in
   Figure 4-4.

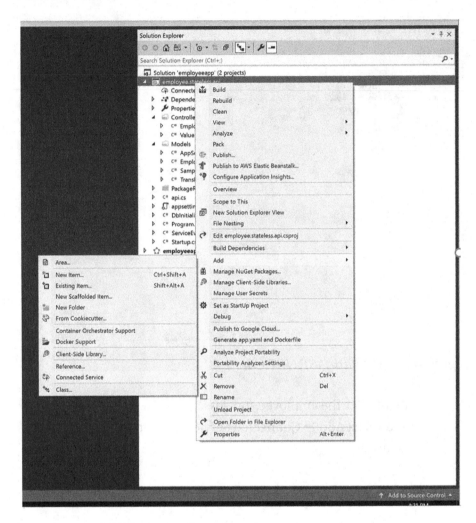

Figure 4-4. *Add connected service*

8.   Choose Monitoring with Application Insights, as
     seen in Figure 4-5.

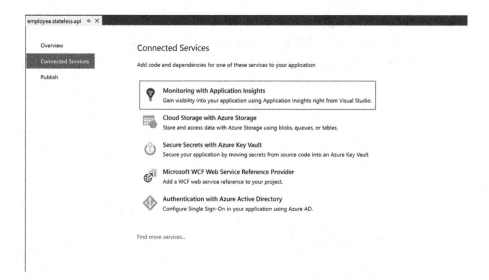

Figure 4-5. *Monitoring with Application Insights*

9.   Click Get Started, as seen in Figure 4-6.

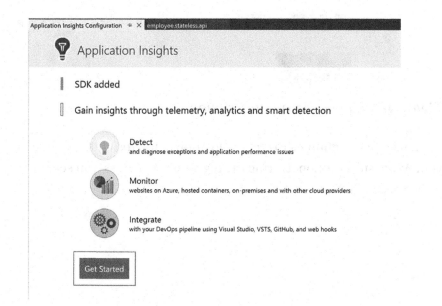

Figure 4-6. *Get started*

10. Choose the right Azure subscription and
Application Insights resource. Once done, click
Register, as seen in Figure 4-7.

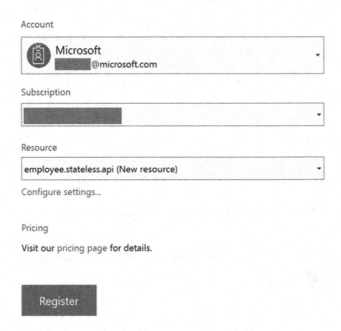

Figure 4-7. *Choose Azure subscription*

It takes a few minutes to create the Application Insights resource in
your Azure subscription. During the registration process, you see the
screen shown in Figure 4-8.

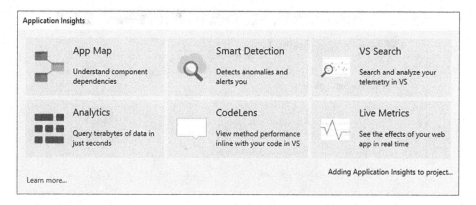

Figure 4-8. *Registration process*

11. Once the Application Insights configuration is
     complete, you see the status as 100%. If you see the
     Add SDK button (as shown in Figure 4-9), click it to
     achieve 100% status, as seen in Figure 4-10.

Figure 4-9. *Add SDK*

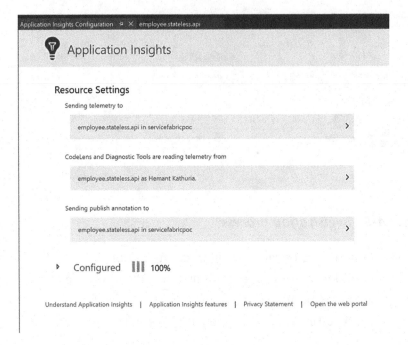

Figure 4-10. *Application Insights SDK installation complete*

12.   To confirm the Application Insights configuration,
       check the instrumentation key in appsettings.json.

13.   Right-click the employee.stateless.api project to add
       dependencies for the following NuGet packages.

       a. Microsoft.EntityFrameworkCore.SqlServer

       b. Microsoft.ApplicationInsights.ServiceFabric.Native

       c. Microsoft.ApplicationInsights.AspNetCore

You are done with the configuration. Now let's add
EmployeeController, which is responsible for performing CRUD
operations on Azure SQL Database.

1. Right-click the employee.stateless.api project and
    add a folder called Models. Add the following
    classes from the sources folder.

    a. AppSettings.cs

    b. Employee.cs

    c. SampleContext.cs

    d. TranslationResponse.cs

2. Right-click the employee.stateless.api project and
    add a file named DbInitializer.cs. Replace that
    content with the following content.

```csharp
using employee.stateless.api.Models;

namespace employee.stateless.api
{
    /// <summary>
    /// Class to initialize database
    /// </summary>
    public class DbInitializer
    {
        private SampleContext _context = null;

        public DbInitializer(SampleContext context)
        {
            _context = context;
        }

        public void Initialize()
        {
            _context.Database.EnsureCreated();
        }
    }
}
```

3. Open Api.cs and replace the contents of the
    CreateServiceInstanceListeners method with the
    following content.

```
return new ServiceInstanceListener[]
            {
                new ServiceInstanceListener(serviceContext =>
                    new KestrelCommunicationListener(service
                    Context, "ServiceEndpoint", (url, listener) =>
                    {
                        ServiceEventSource.Current.
                        ServiceMessage(serviceContext,
                        $"Starting Kestrel on {url}");

                        return new WebHostBuilder()
                                .UseKestrel()
                                //Add the below code to
                                read appsettings.json
                                .ConfigureAppConfiguration(
                                (builderContext, config) =>
                                    {
                                        config.AddJsonFile
                                        ("appsettings.json",
                                        optional: false,
                                        reloadOnChange:
                                        true);
                                })
                                .ConfigureServices(
                                    services => services
                                        .AddSingleton
                                        <StatelessService
                                        Context>(service
                                        Context)
```

```
                                //Make sure the
                                below line exists
                                for application
                                insights
                                integration
                                .AddSingleton
                                <ITelemetry
                                Initializer>
                                ((serviceProvider)
                                => FabricTelemetry
                                Initializer
                                Extension.
                                CreateFabric
                                Telemetry
                                Initializer
                                (serviceContext)))
                        .UseContentRoot(Directory.
                        GetCurrentDirectory())
                        .UseStartup<Startup>()
                        .UseApplicationInsights()
                        .UseServiceFabricIntegration
                        (listener, ServiceFabric
                        IntegrationOptions.None)
                        .UseUrls(url)
                        .Build();
            }))
    };
```

Make sure that you have the following namespaces imported on top of the Api.cs file.

```
using System.Collections.Generic;
using System.Fabric;
using System.IO;
using Microsoft.AspNetCore.Hosting;
using Microsoft.Extensions.DependencyInjection;
using Microsoft.ServiceFabric.Services.Communication.
AspNetCore;
using Microsoft.ServiceFabric.Services.Communication.Runtime;
using Microsoft.ServiceFabric.Services.Runtime;
using Microsoft.Extensions.Configuration;
using Microsoft.ApplicationInsights.Extensibility;
using Microsoft.ApplicationInsights.ServiceFabric;
```

4. Open Startup.cs and replace the contents of the ConfigureServices method with the following content.

```
services.AddDbContext<SampleContext>(options =>

//registring the use of SQL server
            options.UseSqlServer(Configuration.GetConnection
            String("DefaultConnection")));

  services.AddSingleton<DbInitializer>();

  services.AddHttpClient();

  services.Configure<AppSettings>(Configuration.
GetSection("AppSettings"));

  var serviceProvider = services.BuildServiceProvider();
```

```
var dbInitializer = serviceProvider.GetRequiredService
<DbInitializer>();

dbInitializer.Initialize();
```

```
services.AddMvc().SetCompatibilityVersion(CompatibilityVersion.
Version_2_2);
```

5. Right-click the controller folder in the employee.
   stateless.api project and add a controller called
   EmployeeController.cs. Replace that content with
   the following content.

```
using System.Collections.Generic;
using System.Linq;
using System.Net.Http;
using System.Text;
using System.Threading.Tasks;
using employee.stateless.api.Models;
using Microsoft.AspNetCore.Mvc;
using Microsoft.EntityFrameworkCore;
using Microsoft.Extensions.Options;
using Newtonsoft.Json;
using Newtonsoft.Json.Linq;

namespace employee.stateless.api.Controllers
{
    [Route("api/[controller]")]
    [ApiController]
    public class EmployeeController : ControllerBase
    {
        /// <summary>
        /// Context
        /// </summary>
```

```csharp
private SampleContext _context = null;

private HttpClient _httpClient = null;

private AppSettings _appSettings = null;

/// <summary>
/// Employee Controller
/// </summary>
/// <param name="context"></param>
public EmployeeController(SampleContext
context, IHttpClientFactory httpClientFactory,
IOptionsMonitor<AppSettings> appSettings)
{
    _context = context;

    _appSettings = appSettings.CurrentValue;

    _httpClient = httpClientFactory.CreateClient();
}

/// <summary>
/// Returns all the employees
/// </summary>
/// <returns></returns>
[HttpGet]
public async Task<ActionResult<List<Employee>>>
GetAll()
{
    List<Employee> employeeList = await _context.
    Employees.ToListAsync<Employee>();

    return new OkObjectResult(employeeList);
}
```

```csharp
/// <summary>
/// Returns an employee based on id
/// </summary>
/// <param name="id"></param>
/// <returns></returns>
[HttpGet("{id}")]
public async Task<ActionResult<Employee>>
GetEmployee(int id)
{
    Employee employee = await _context.Employees.
    Where(e => e.Id.Equals(id)).FirstOrDefaultAsync();

    return new OkObjectResult(employee);
}

/// <summary>
/// Creates an emaployee
/// </summary>
/// <param name="employee"></param>
/// <returns></returns>
[HttpPost]
public async Task<ActionResult> Post(Employee employee)
{
    employee.NativeLanguageName = await
    GetTranslatedText(employee.FirstName);

    await _context.Employees.AddAsync(employee);

    await _context.SaveChangesAsync();

    return new OkResult();
}
```

```csharp
/// <summary>
/// Deletes the employee based on id
/// </summary>
/// <param name="id"></param>
/// <returns></returns>
[HttpDelete("{id}")]
public async Task<ActionResult> Delete(int id)
{
    Employee employee = await _context.Employees.
    Where(e => e.Id.Equals(id)).FirstOrDefaultAsync();

    if (employee == null)
    {
        return new NotFoundResult();

    }else
    {
        _context.Employees.Remove(employee);

        await _context.SaveChangesAsync();

        return new OkResult();
    }
}
/// <summary>
/// Gets the name in hindi
/// </summary>
/// <param name="name"></param>
/// <returns></returns>
private async Task<string> GetTranslatedText(string
name)
{
    System.Object[] body = new System.Object[] { new {
    Text = name } };
```

```csharp
        var requestBody = JsonConvert.SerializeObject(body);

        StringContent content = new StringContent
        (requestBody, Encoding.UTF8, "application/json");

        _httpClient.DefaultRequestHeaders.Add("Ocp-Apim-
        Subscription-Key", _appSettings.AccessKey);

        var result =  await _httpClient.PostAsync
        ($"{_appSettings.TranslationApiUrl}/translate?api-
        version=3.0&to=hi", content);

        result.EnsureSuccessStatusCode();

        string translatedJson = await result.Content.
        ReadAsStringAsync();

        TranslationResponse response = Newtonsoft.Json.
        JsonConvert.DeserializeObject<TranslationResponse>
        (JArray.Parse(translatedJson)[0].ToString());

        return response.translations[0].text;

      }
    }
}
```

6. Open AppSettings.json and make sure that the content looks similar to your connection strings.

```json
{
  "ConnectionStrings": {
    "DefaultConnection": "Server=<<YOUR_SERVER>>;Database=
    <<YOUR_DATABASE>>;User ID=<<USER_ID>>;Password=<<PASSWORD>>
    ;Trusted_Connection=False;Encrypt=True;MultipleActiveResult
    Sets=True;"
  },
```

```
"AppSettings": {
  "TranslationApiUrl": "https://api.cognitive.
  microsofttranslator.com",
  "AccessKey": "<<YOUR ACCESS KEY TO TRANSLATION API>>"
},
"Logging": {
  "LogLevel": {
    "Default": "Warning"
  }
},
"AllowedHosts": "*",
"ApplicationInsights": {
  "InstrumentationKey": "<<YOUR INSTRUMENTATION KEY OF YOUR
  APP INSIGHTS RESOURCE>>"
}
}
```

7. Run the application against your local Service Fabric cluster to perform a simple test. Since it is an API, you can test it using tools like Postman or Fiddler. Figure 4-11 is a sample API call from Fiddler. Execute the API multiple times so that you can easily monitor the traffic in Application Insights. Now let's deploy the application on a Service Fabric cluster.

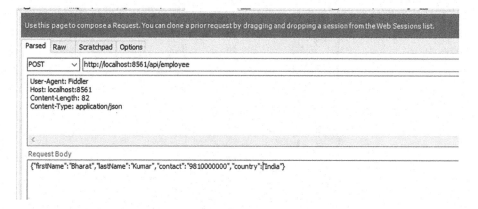

Figure 4-11. *Sample Fiddler call*

8. Make sure that you install the .pfx certificate on your desktop (it was created along with the Service Fabric cluster on Azure). You can download the certificate from the Azure portal, as seen in Figure 4-12. This is required to publish your app on an Azure Service Fabric cluster.

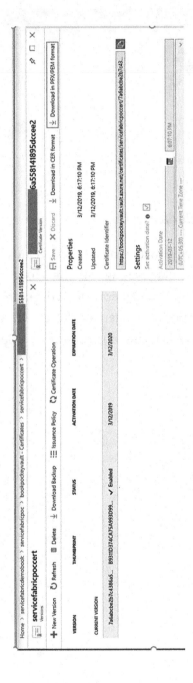

Figure 4-12. Download certificate

9. Make sure that the port of your service is open in the
    load balancer for the Azure Service Fabric cluster, as
    seen in Figure 4-13. You can find your service port
    in ServiceManifest.xml in the employee.stateless.api
    project.

```
<Endpoint Protocol="http" Name="ServiceEndpoint" Type="Input"
Port="80" />
```

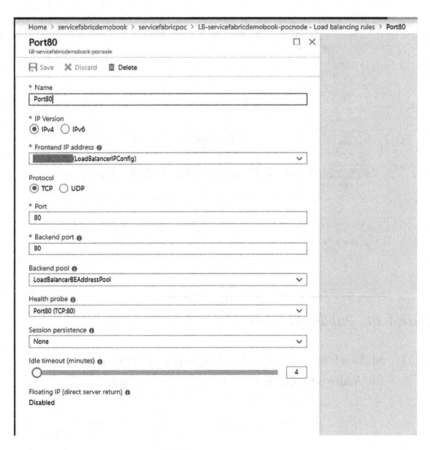

Figure 4-13. *Service port configuration*

10.   Right-click the EmployeeApp project and choose
      Publish. Select Cloud.xml in the target profile, as
      seen in Figure 4-14. Make sure to specify the right
      certificate thumbprint, store location, and store
      name.

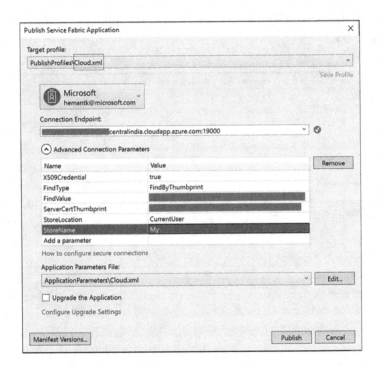

Figure 4-14.   Publish to Azure

11.   Make a few calls to the Post Employee API hosted on
      the Azure Service Fabric cluster, as seen in Figure 4-15.

Figure 4-15. *Fiddler call to post employee API*

At the time of creating an employee record, we are inserting a record in Azure SQL and invoking the Azure Translator Text API. Let's look at how informative Azure Application Insights is.

1. Log in to the Azure portal, select the Application Insights resource, and click Search, as seen in Figure 4-16.

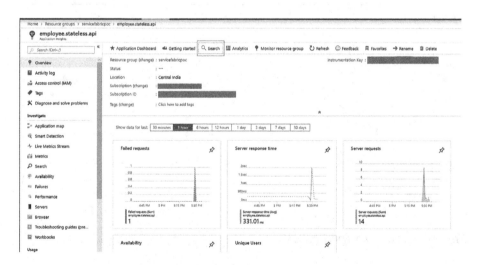

Figure 4-16. *Application Insights view*

2. Click **Click here to see all data in the last 24 hours**, as seen in Figure 4-17. Please note that it may take a few minutes before you can monitor the traffic on the Azure portal.

Figure 4-17. *Last 24 hours details*

3. Once you see the requests, click **Grouped results**, as seen in Figure 4-18.

Figure 4-18. *Grouped results*

It automatically recognizes the remote calls to the database and the Translator Text API. You can click further to see information like SQL query, time taken by query, Service Fabric node details, and so forth, as seen in Figure 4-19.

Figure 4-19. *SQL query details*

It should be very clear how easily you can monitor a Service Fabric application by using Application Insights. It automatically detects remote SQL and HTTP dependencies, which is a great feature for optimizing application performance.

Cluster Monitoring

One of the salient features of an Azure Service Fabric cluster is making applications resilient to hardware failures. For example, if Service Fabric system services are having issues in deploying workloads, or services are not able to enforce placement rules, Service Fabric provides diagnostic logs to monitor these scenarios. Service Fabric exposes various structured platform events for efficient diagnosis.

On Azure, for windows clusters, it's suggested to use Diagnostic Agents and Azure Monitor Logs. Azure Monitor Logs is also suggested for Linux workloads, but with a different configuration.

Diagnostic Agents

The Windows Azure Diagnostic extension allows you to collect all the logs from all the cluster nodes to a central location. The central location can be Azure Storage, and it can send the logs to Azure Application Insights or Event Hubs.

Diagnostic agents can be deployed through the Azure portal when creating a Service Fabric cluster. You can also use the Resource Manager template to add a diagnostic agent to an existing Service Fabric cluster, if it was not added when creating the cluster. Figure 4-20 shows the diagnostics options when creating a Service Fabric cluster.

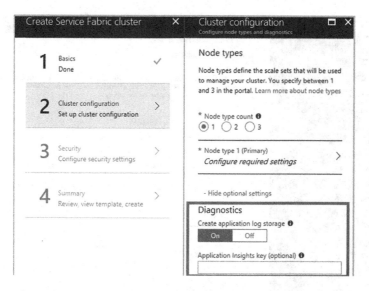

Figure 4-20. Configure diagnostics agent

Azure Monitor Logs

Microsoft recommends using Azure Monitor Logs to monitor cluster-level events in a Service Fabric cluster. To use this option, the diagnostic logs for the Service Fabric cluster must be enabled.

Setting up Azure Monitor Logs is done through Azure Resource Manager, PowerShell, or Azure Marketplace. Here we follow the Azure Marketplace route because it's user-friendly and easy to understand.

1. Select New in the left navigation menu of the Azure portal.

2. Search for Service Fabric Analytics. Select the resource that appears.

3. Select Create, as seen in Figure 4-21.

Figure 4-21. *Create Service Fabric analytics*

1. Create a new Log Analytics workspace, as seen in Figure 4-22. Once it is created, you need to connect it to your Azure Service Fabric cluster.

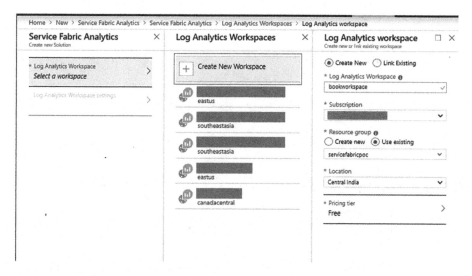

Figure 4-22. *Create log analytics*

2. Go to the resource group in which you created the Service Fabric analytics solution. Select ServiceFabric<nameOfWorkspace> and go to its Overview page.

3. Select Storage Account Logs under the Workspace Data Sources option.

4. Click Add, as seen in Figure 4-23.

Figure 4-23. *Add storage account logs*

5. Choose the storage account created with the Service
    Fabric cluster. The default name for the Service
    Fabric cluster storage account starts with sfdg.

6. Select Service Fabric Events as the data type.

7. Make sure that the source is set to
    WADServiceFabric*EventTable, as seen in Figure 4-24.

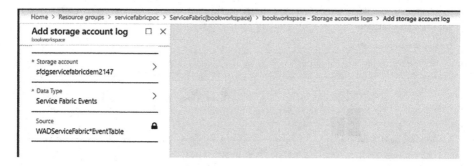

Figure 4-24. *WADServiceFabric*EventTable*

Once done, on the Overview page, you see a summary of Service Fabric events. Please note that it may take 10 to 15 minutes for data to appear in this view, as seen in Figure 4-25.

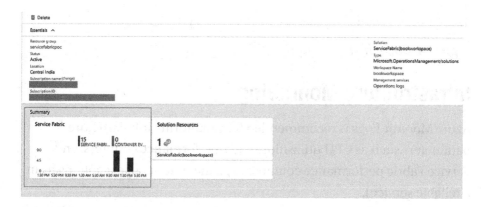

Figure 4-25. *Overview of Service Fabric events*

8. Click the Service Fabric tile to see more reported information about the cluster events, as seen in Figure 4-26.

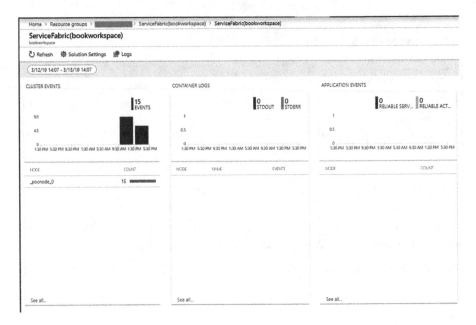

Figure 4-26. *Cluster event details*

Infrastructure Monitoring

Azure Monitor Logs is recommended for monitoring infrastructure parameters such as CPU utilization, .NET performance counters, and Service Fabric performance counters (e.g., the number of exceptions from a reliable service).

To get the infrastructure logs, you are required to add a Log Analytics agent as a virtual machine scale set extension to the Azure Service Fabric cluster.

Follow these steps to do this.

1. Go to the resource group in which you created the Service Fabric Analytics solution. Select ServiceFabric<nameOfWorkspace> and go to its Overview page. Select Log Analytics Workspace and click Advanced Settings under Settings.

2. Select Windows Servers. Make a note of the
    workspace ID and primary key, as seen in
    Figure 4-27.

Figure 4-27. *Windows Server details*

3. Open the Cloud Shell from the Azure portal to
    run the command in the next step. The option is
    available in the top-right corner of the Azure portal,
    as seen in Figure 4-28.

Figure 4-28. *Cloud shell*

4. Execute the following command to add the
    monitoring agent.

```
az vmss extension set --name MicrosoftMonitoringAgent
--publisher Microsoft.EnterpriseCloud.Monitoring --resource-
group <nameOfResourceGroup> --vmss-name <nameOfNodeType>
--settings "{'workspaceId':'<Log AnalyticsworkspaceId>'}"
--protected-settings "{'workspaceKey':'<Log
AnalyticsworkspaceKey>'}"
```

5. Replace the workspace ID and workspace key
    collected from the previous step. nameOfNodeType
    is the name of the virtual machine scale set resource
    that was automatically created with your Service
    Fabric cluster. This command takes about 15
    minutes to add the log analytics agents on all the
    scale set nodes.

6. Go to the resource group in which you created
    the Service Fabric Analytics solution. Select
    ServiceFabric<nameOfWorkspace> and go to its
    Overview page. Select Log Analytics Workspace and
    click Advanced Settings under Settings.

7. Choose Data and Windows Performance Counters.
    Click **Add the selected performance counters.** (For
    the purpose of this sample exercise, we selected the
    default performance counters, but you can choose
    custom performance counters in the real-world
    applications, as seen in Figure 4-29.)

Figure 4-29. *Windows performance counters*

8.   Click Save.

9.   Go to the resource group in which you created
     the Service Fabric Analytics solution. Select
     ServiceFabric<nameOfWorkspace> and go to its
     Overview page. Click the tile for the Summary of
     Service Fabric events.

10. You see data for the selected performance counters,
     like disk usage (MB). Click the chart to get more
     information, as seen in Figure 4-30. Please note that
     it takes time to reflect data in this section.

Disk usage (MB)

|| pocnode000000

FreeDiskSpace
100k
80k
60k
40k
20k

12:00 PM   6:00 PM   12:00 AM   6:00 AM

Figure 4-30. *Disk usage details*

Summary

In this chapter, you learned how to monitor a Service Fabric cluster and the applications deployed on it. We covered application monitoring, cluster monitoring, and infrastructure monitoring. Application Insights is a very effective approach to monitoring deployed applications because it effectively monitors remote HTTP and database calls with no additional effort. We also covered how you can use Diagnostic Agents and Azure Monitor Logs to monitor a Service Fabric cluster and infrastructure.

CHAPTER 5

Azure Kubernetes Service

Previous chapters discussed the benefits and various challenges of microservices applications. By now, it's quite clear that orchestrators are the backbone of the microservices ecosystem. We also discussed Azure Service Fabric and the various options to deploy microservices applications. Azure Kubernetes Service (AKS) is a vast subject; however, we will cover the fundamental details required to deploy microservices applications. There are practical scenarios along with hands-on code to give you a fair idea of how easy it is to adopt AKS.

Introduction to Kubernetes

Before we get into AKS, let's go over some of the Kubernetes platform's background. Kubernetes (a.k.a. K8s) is an extensible, open source container orchestration system for automating application deployment, scaling, and management. It aims to provide a "platform for automating deployment, scaling, and operations of application containers across clusters of hosts." It works with a range of container tools, including Docker, and has a large, rapidly growing ecosystem.

As an open platform, Kubernetes allows you to build applications with your preferred programming language and operating system. It can be

© Harsh Chawla and Hemant Kathuria 2019
H. Chawla and H. Kathuria, *Building Microservices Applications on Microsoft Azure*,
https://doi.org/10.1007/978-1-4842-4828-7_5

seamlessly integrated with continuous integration and continuous delivery (CI/CD) to schedule and deploy releases. Adoption of this platform is growing rapidly, and many cloud providers offer the Kubernetes platform on IaaS and PaaS. AKS is Microsoft's PaaS service for Kubernetes.

Let's now discuss Kubernetes' architecture and important components.

Kubernetes Cluster Architecture

Kubernetes follows a master-slave architecture. As shown in Figure 5-1, it's divided into two parts.

- **Master**. The main controlling unit of the cluster, it provides the core services and orchestration of application workloads.

- **Node**. The worker nodes, or minions, within the cluster that are responsible for running application workloads.

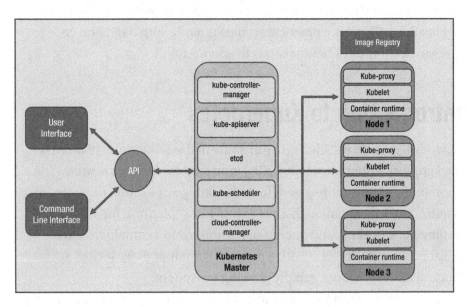

Figure 5-1. *AKS architecture*

Kubernetes Master

In Kubernetes, the master is the main controlling unit for the cluster. It manages all the workloads and directs communication across the system. Figure 5-1 shows that the K8s master includes the following core Kubernetes components.

- **kube-controller-manager**. The controller manager oversees the collection of smaller controllers. Controllers perform actions such as replicating pods and handling node operations, such as maintaining the correct number of nodes, maintaining high availability, and so forth. It moves pods to other nodes if any underlying node fails.

- **cloud-controller-manager**. The controller manager enables the cloud provider to integrate with Kubernetes. Kubernetes has plugins for cloud providers to integrate and customize the platform. AKS uses this component to build a PaaS service on Azure.

- **kube-apiserver**. The API server exposes underlying Kubernetes APIs to interact with management tools, such as kubectl or the Kubernetes dashboard.

- **etcd**. A key/value pair database that maintains the state of the Kubernetes cluster and configuration. It is a highly available key/value store within Kubernetes.

- **kube-scheduler**. The scheduler is responsible for allocating the compute and storage required to create and scale containerized applications. The scheduler determines which nodes can accommodate the workload, and runs it on them.

Interaction with the cluster master is facilitated by the Kubernetes APIs or dashboard, or CLI-based kubectl.

Kubernetes Nodes

Kubernetes nodes run your applications and supporting services within the Kubernetes cluster by running the following components and container runtime (as shown in Figure 5-1).

- The kubelet is the Kubernetes agent that runs on each node in the cluster and processes the orchestration requests from the cluster master. It ensures that the containers described in the configuration are up and running.

- kube-proxy manages networking operations on each node. This includes routing network traffic and managing IP addresses for services and pods.

- The container runtime is responsible for running the containers on the Kubernetes cluster. It allows containerized applications to run and interact with resources, such as virtual networks and storage. There are multiple container runtimes supported on Kubernetes (e.g., Docker, rklet, etc.).

A Kubernetes cluster can have multiple nodes, and each node translates into an Azure virtual machine and hosts the services. Depending on your application's resource needs, you can choose the VM size to make sure that you have enough CPU, memory, and storage.

Note   In AKS, the VM images are based on Ubuntu Linux. If you need to support any other OS, you can leverage aks-engine (tooling to quickly bootstrap Kubernetes clusters on Azure) and set up your own Kubernetes cluster.

Kubernetes Node Pools

A Kubernetes cluster allows you to group nodes with the same configuration into node pools. A cluster can have more than one node pool. When creating a cluster in Azure, a default pool is automatically created for you.

Pods

A pod is a logical resource that represents the container hosting the instance of your application. Although, it is supported to run multiples containers in a pod, we haven't faced many scenarios where it was needed.

When creating a pod, you can specify the required resource limits, and kube-scheduler tries to run the pod on the node where the required resources are available.

At runtime, pods are scaled by creating replica sets, and it makes sure that the required number of pods are always running in a Kubernetes deployment.

Deployment

A deployment in a Kubernetes cluster is managed by a Kubernetes deployment controller, which internally uses kube-scheduler. To define a Kubernetes deployment, a manifest file is maintained in YAML format.

With the help of replica sets, the required number of pods are always running in a Kubernetes deployment. The manifest file mentions the required number of replicas. The manifest file also includes the container image, ports that need to be opened in the container, attached storage, and so forth.

The following is a sample YAML file.

```
apiVersion: apps/v1
kind: Deployment
metadata:
 name: nginx0-deployment
 labels:
 app: nginx0-deployment
spec:
 replicas: 2
 selector:
 matchLabels:
 app: nginx0
 template:
 metadata:
 labels:
 app: nginx0
 spec:
 containers:
 - name: nginx
 image: nginx:1.7.9
 ports:
 - containerPort: 80
```

If the application deployed on a cluster are stateless, the deployment controller runs an instance of an application on any of the available nodes.

If the deployed applications are stateful, Kubernetes uses the following resources.

StatefulSets

A StatefulSet maintains the sticky identity of the pods. It is used if the application needs persistence storage, unique network identifiers, and ordered deployment and scaling. The application in YAML is defined using kind: StatefulSet.

DemonSets

If a pod needs to run on all the available nodes, DaemonSet makes sure that all nodes are running a copy of a pod. The pods scheduled by DaemonSet are started before the pods that are scheduled by deployment or StatefulSet controllers. The application in YAML is defined using kind: DaemonSet.

Namespaces

A namespace allows you to logically divide a Kubernetes cluster between multiple users to control creating and managing the resources. It is needed in an environment where users are from multiple teams and a cluster is shared between them. Namespaces provide scope for names (i.e., a resource name should be unique within a namespace). You should not use namespaces for a cluster with few or limited users.

What Is Azure Kubernetes Service?

AKS is a managed Kubernetes service on the Microsoft Azure platform. Since it's a managed service, Azure manages all the ongoing Kubernetes operations and maintains the platform by upgrading and scaling resources on demand. It eases the hosting and management of a Kubernetes environment and enables the end user to deploy containerized applications without spending time setting up Kubernetes clusters.

In short, Kubernetes masters are managed by Azure and end users manage the agent nodes.

The following are the key advantages of using AKS and explains how the power of the cloud reinforces the capabilities of Kubernetes.

- Access to Azure enterprise-class features, including identity management, integrated monitoring, and networking.

- Reduces complexity for end users by offloading operational overhead to the Azure platform.

- Azure automatically handles Kubernetes cluster monitoring and maintenance.

- The platform can be quickly built or destroyed, as needed.

While there are a lot of advantages for using AKS, there are also limitations; for example, Windows containers are not supported (as of the time of writing this book). Moreover, applications can be deployed as only containers, unlike with Service Fabric, where there are more options, such as stateless/stateful services, and so forth.

Let's now move to configuring AKS and building a containerized application. These practical labs will give you a lot of clarity on how the platform can be utilized. We'll start with the basics of setting up the IDE.

AKS Development Tools

Developing a Kubernetes application can be challenging. You need Docker and Kubernetes configuration files. You need to figure out how to test your application locally and to interact with other dependent services. You might need to handle developing and testing on multiple services at once and with a team of developers.

Azure Dev Spaces helps you develop, deploy, and debug Kubernetes applications directly in AKS. Azure Dev Spaces also allows a team to share a dev space. Sharing a dev space across a team allows individual team members to develop in isolation without having to replicate or mock up dependencies or other applications in the cluster.

Azure Dev Spaces creates and uses a configuration file for deploying, running, and debugging your Kubernetes applications in AKS. This configuration file resides within your application's code and can be added to your version control system.

Set up AKS and Development Tools for Labs

So far in this chapter, we have covered Azure Kubernetes Service and programming tools, and you learned how to install it on the cloud. Now let's create a few samples to better understand how to develop and deploy applications. First, let's create an AKS cluster instance on Azure.

Create an Azure Kubernetes Service Cluster

Please follow the steps covered in this section to configure an AKS cluster. This setup is used to deploy a containerized application on this platform.

1. Sign in to the Azure portal.

2. Select + Create a resource ➤ Kubernetes Service.

3. Enter the subscription, resource group, Kubernetes cluster name, region, Kubernetes version, and DNS name prefix, as shown in Figure 5-2.

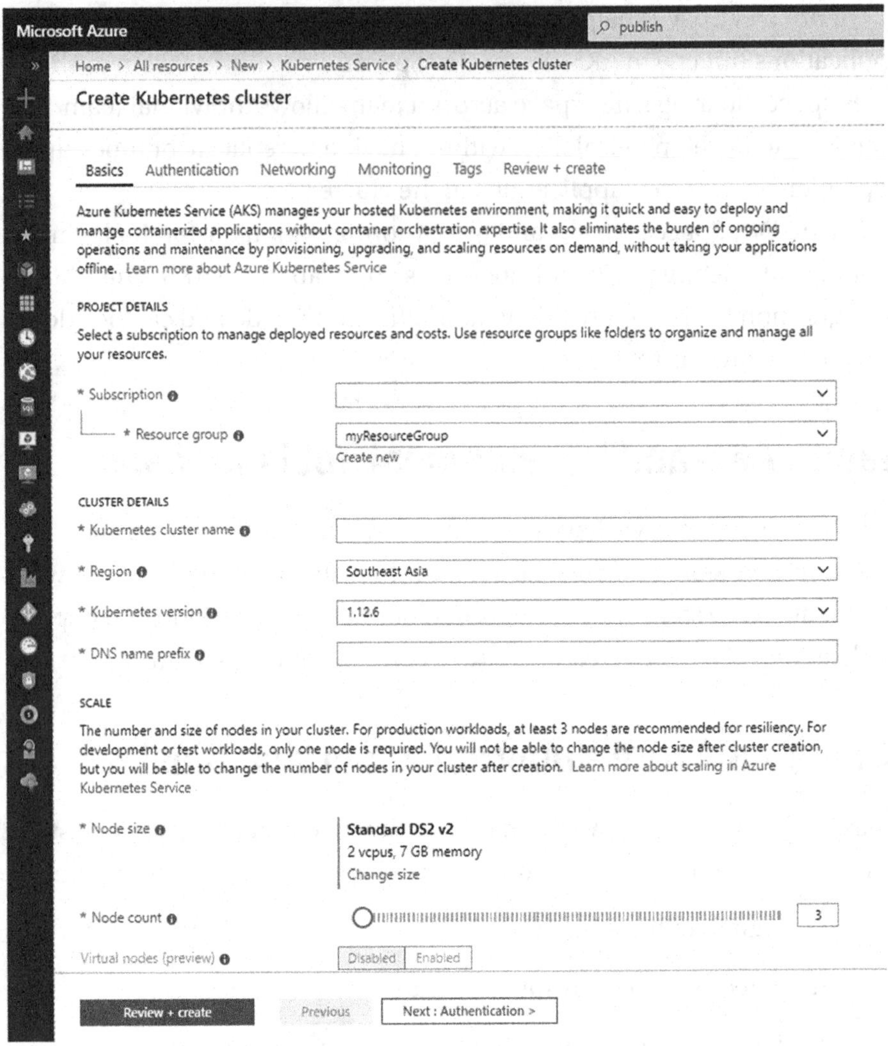

Figure 5-2. *Set up AKS cluster*

4. Click Review + create.

5. Click Create.

This creates the different resources that are needed to support the AKS service (e.g., virtual machines for the agent pool, networking stuff, and Kubernetes service).

Enable Azure Dev Spaces on an AKS Cluster

Navigate to your AKS cluster in the Azure portal and click Dev Spaces. Select Yes for Enable Dev Spaces and click Save, as shown in Figure 5-3.

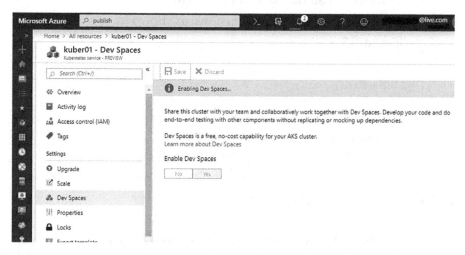

Figure 5-3. *Enable Dev Spaces on AKS*

Configure Visual Studio to Work with an Azure Kubernetes Service Cluster

In the preceding sections, an AKS cluster and dev spaces were set up successfully. Now, let's set up extension for Kubernetes in Visual Studio.

1. Open Visual Studio 2017 ➤ Tools ➤ Extensions and Updates.

2. Search for Kubernetes Tools, as shown in Figure 5-4.

161

Figure 5-4. *Install Kubernetes extensions for Visual Studio*

3. Click the download button, which queues the installer.

4. Close Visual Studio 2017 so that the installer is launched to modify the extensions, as shown in Figure 5-5.

Figure 5-5. *Install Kubernetes extension*

5. Create an ASP.NET Core web application (Model-View-Controller) with .NET Core 2.0.

6. Once the extension is installed, you should see Azure Dev Spaces in the menu, as shown in Figure 5-6.

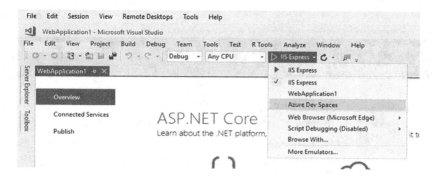

Figure 5-6. *Azure Dev Spaces*

Now you are good to go ahead with the lab.

Configure Visual Studio Code to Work with an Azure Kubernetes Service Cluster

Once Azure Dev Spaces is installed, let's configure the Visual Studio Code as well. It's a new product that is lightweight and available for Linux or Mac users. It's primarily for front-end developers; it is not for managing an entire project/debugging/integration with source control and so forth. Let's install Kubernetes extensions in Visual Studio Code.

1. Open a browser and enter **https://marketplace. visualstudio.com/VSCode**.

2. Click the Search button and enter **Azure Dev Spaces** (see Figure 5-7).

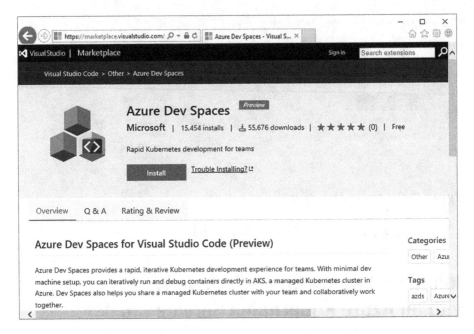

Figure 5-7. *Azure Dev Spaces for Visual Studio Code*

3. Click the Install button to install, which opens a confirmation dialog.

4. Once you confirm by clicking the Allow button, the extensions installation windows open in Visual Studio Code.

5. Click install on the bottom right as shown in Figure 5-8 and this extension would be installed.

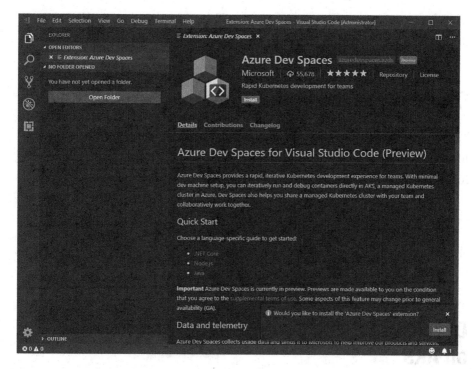

Figure 5-8. *Install Dev Spaces*

6. Install Azure CLI from https://docs.microsoft.
    com/en-us/cli/azure/install-azure-
    cli?view=azure-cli-latest.

7. Once Azure CLI is installed, run the following Azure
    CLI command (see Figure 5-9), which installs the
    AZDS utility in the background. First, you have to
    log in with the az login command.

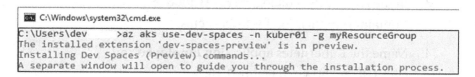

Figure 5-9. *Install AZDS utility*

We are done installing AZDS, Azure CLI, and extensions in Visual Studio Code.

Deploy Application on AKS

The development tools Visual Studio and Visual Studio Code are set up for application development. Let's deploy two very simple applications as follows.

- Demonstrate developing the ASP.NET Core web app and deploy it on AKS.

- Demonstrate developing Node.js using Visual Studio Code and deploy it on AKS.

Develop ASP.NET Core Web App and Deploy on AKS

In this section, we create an ASP.Net Core web API in Visual Studio. It will deploy a web API container on AKS.

Create an ASP.NET Core Web API

Let's create a web API.

1. Launch Visual Studio as an administrator.

2. Create a project with File ➤ New ➤ Project.

3. In the New Project dialog, choose Cloud ➤ Container Application for Kubernetes.

4. Name the Kubernetes application **Employee** and click OK, as shown in Figure 5-10.

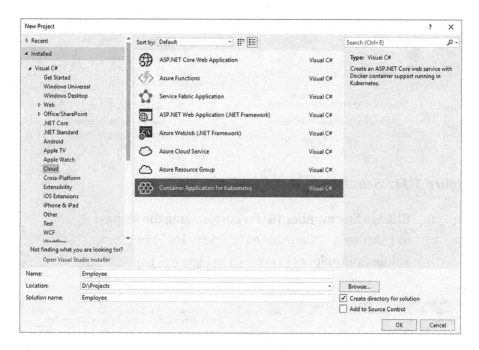

Figure 5-10. *Create a container application*

5.   Choose the API from the next dialog box and
     click OK.

6.   Once the project is created, there is a confirmation
     box to make it a publicly accessible endpoint.
     Click Yes.

7.   Make sure that Azure Dev Spaces is selected in the
     menu, as shown in Figure 5-11.

Figure 5-11. *Select Azure Dev Spaces*

8. Change line number 16 (for customizing the strings) to { "Azure", "Kubernetes", "Service") in ValuesController.cs (as shown in Figure 5-12).

```
ValuesController.cs*  ⊕ ✕   azds.yaml        Employee
🗐 Employee                              ▾  ❖ Employee.Controllers.ValuesController
    1   ⊟using System;
    2    │using System.Collections.Generic;
    3    │using System.Linq;
    4    │using System.Threading.Tasks;
    5    └using Microsoft.AspNetCore.Mvc;
    6
    7   ⊟namespace Employee.Controllers
    8    │{
    9    │    [Route("api/[controller]")]
             0 references
   10   ⊟    public class ValuesController : Controller
   11    │    {
   12    │        // GET api/values
   13    │        [HttpGet]
                   0 references | 0 requests | 0 exceptions
   14   ⊟        public IEnumerable<string> Get()
   15    │        {
   16/   │            return new string[] { "Azure", "Kubernetes", "Service" };
   17    │        }
```

Figure 5-12. *Change the output string*

9. Hit F5 to run the application.

Azure Dev Spaces confirms the Microsoft account, subscription and cluster information, as shown in Figure 5-13.

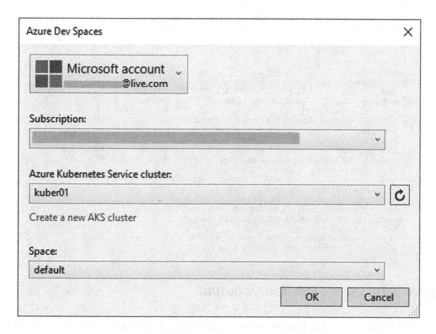

Figure 5-13. Azure Dev Spaces credentials confirmation page

10. Hit OK. This deploys your application to AKS on
     Azure.

11. Check Output from Azure Dev Spaces(as shown in
     Figure 5-14), which mainly regards the creation of a
     container from the Docker file and hosting it on AKS
     with a public endpoint.

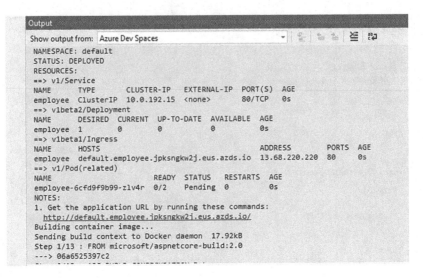

Figure 5-14. *Azure Dev Spaces output*

In a browser or cURL command, you can see the output from a public endpoint.

Develop Node.js Using Visual Studio Code and Deploy It on AKS

Let's deploy a Node.js application in Visual Studio Code.

Create a Node.js API

Follow these steps to create a Node.js API application.

1. In Windows Explorer, create a project folder named myApp in your preferred folder.

2. Launch Visual Studio Code as an administrator.

3. Open the project folder with the File ➤ Open Folder
   menu.

4. Create a new file by selecting File ➤ New File. Name
   the file **server.js** and add the following code to it.

```
var express = require('express');
var app = express();
app.use(express.static(__dirname + '/public'));

app.get('/', function (req, res) {
    res.sendFile(__dirname + '/public/index.html');
});

app.get('/api', function (req, res) {
    res.send('Hello from webfrontend');
});

var port = process.env.PORT || 80;
var server = app.listen(port, function () {
    console.log('Listening on port ' + port);
});

process.on("SIGINT", () => {
    process.exit(130 /* 128 + SIGINT */);
});

process.on("SIGTERM", () => {
    console.log("Terminating...");
    server.close();
});
```

5. Create a new file by selecting File ➤ New File. Name
    the file **package.json** and add the following code to it.

```
{
  "name": "webfrontend",
  "version": "0.1.0",
  "devDependencies": {
    "nodemon": "^1.18.10"
  },
  "dependencies": {
    "express": "^4.16.2",
    "request": "2.83.0"
  },
  "main": "server.js",
  "scripts": {
    "start": "node server.js"
  }
}
```

6. From Windows Explorer within the myApp folder,
    create another folder and name it **public**.

7. Create a new file in the public folder by selecting
    File ➤ New File. Name the file **index.html** and add
    the following code to it.

```
<!doctype html>
<html ng-app="myApp">
<head>
    <script src="https://ajax.googleapis.com/ajax/libs/
    angularjs/1.5.3/angular.min.js"></script>
    <script src="https://cdnjs.cloudflare.com/ajax/
    libs/angular.js/1.5.3/angular-route.js"></script>
```

```html
<script src="app.js"></script>
<link rel="stylesheet" href="app.css">
<link href="https://maxcdn.bootstrapcdn.
com/bootstrap/3.3.6/css/bootstrap.min.css"
rel="stylesheet" integrity="sha384-1q8mTJOASx8j1Au
+a5WDVnPi2lkFfwwEAa8hDDdjZlpLegxhjVME1fgjWPGmkzs7"
crossorigin="anonymous">
<!-- Uncomment the next line -->
<!-- <meta name="viewport" content="width=device-
width, initial-scale=1"> -->
</head>
<body style="margin-left:10px; margin-right:10px;">
    <div ng-controller="MainController">
        <h2>Server Says</h2>
        <div class="row">
            <div class="col-xs-8 col-md-10">
                <div ng-repeat="message in messages
                track by $index">
                    <span class="message">{{message}}</
                    span>
                </div>
            </div>
            <div class="col-xs-4 col-md-2">
                <button class="btn btn-primary"
                ng-click="sayHelloToServer()">Say It
                Again</button>
            </div>
        </div>
    </div>
</body>
</html>
```

8. Create a new file in the public folder by selecting
    File ➤ New File. Name the file **app.js** and add the
    following code to it.

```
var app = angular.module('myApp', ['ngRoute']);

app.controller('MainController', function($scope, $http) {

    $scope.messages = [];
    $scope.sayHelloToServer = function() {
        $http.get("/api?_=" + Date.now()).
        then(function(response) {
            $scope.messages.push(response.data);
        });
    };

    $scope.sayHelloToServer();
});
```

9. Finally, create a new file in public folder by selecting
    File ➤ New File. Name the file **app.css** and add the
    following code to it.

```
.message {
    font-family: Courier New, Courier, monospace;
    font-weight: bold;
}
```

Note   This project can also be created with the Express application
if you have installed Node.js and npm with the following commands.

---

1. Install the Express Generator by running the
    following from a terminal.

    ```
    npm install -g express-generator
    ```

2. Scaffold a new Express application called
    myExpressApp by running the following.

    ```
    express myExpressApp
    ```

3. Open a new Express application by opening a folder
    in Visual Studio Code.

4. Open the Command Palette in Visual Studio Code.

5. Click View and then Command Palette, or press
    Ctrl+Shift+P.

6. Enter Azure Dev Spaces and click Azure Dev Spaces.

7. Prepare the configuration files for Azure Dev Spaces
    (as shown in Figure 5-15).

Figure 5-15. *Prepare Azure Dev Space Environment*

8. Click the Debug icon on the left, and then click
    Launch Server (AZDS) (as shown in Figure 5-16).

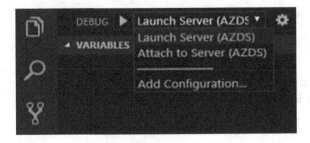

Figure 5-16. *Select AZDS in the debugger*

The Debug console shows the log output.

```
> Executing task: C:\Program Files\Microsoft SDKs\Azure\Azure
Dev Spaces CLI (Preview)\azds.exe up --port=50521:9229 --await-
exec --keep-alive <

Synchronizing files...4s
Using dev space 'new01' with target 'kuber01'
Installing Helm chart...2s
Waiting for container image build...29s
Building container image...
Step 1/8 : FROM node:lts
Step 2/8 : ENV PORT 80
Step 3/8 : EXPOSE 80
Step 4/8 : WORKDIR /app
Step 5/8 : COPY package.json .
Step 6/8 : RUN npm install
Step 7/8 : COPY . .
Step 8/8 : CMD ["npm", "start"]
Built container image in 45s
Waiting for container...52s
Service 'myapp' port 80 (http) is available via port forwarding
at http://localhost:50764

Terminal will be reused by tasks, press any key to close it.
```

Summary

In this chapter, we covered the most important parts of AKS, including its functionalities and benefits. AKS eases managing a microservices ecosystem. It takes away the burden of setting up, configuring, managing, and monitoring a Kubernetes cluster. Developers can simply run their containerized applications on agent nodes and harness the power of autoscaling, high availability, and so forth, by using AKS.

CHAPTER 6

Monitoring Azure Kubernetes Service

Chapter 5 was an in-depth discussion of Azure Kubernetes Service and its core concepts, including masters, agent nodes, application scaling, supported programming models and the tools available to work with Visual Studio and so forth. AKS is a managed service that provides native capabilities to monitor an entire infrastructure. In this chapter, we will explore the various options to monitor the AKS ecosystem.

Monitoring

The dictionary meaning of *monitoring* is to observe and check the progress or quality of something over a period of time. It's an important part of an administrator's job to ensure that the services are up and running. AKS has four important components that should be monitored.

- Clusters
- Nodes
- Controllers
- Containers

© Harsh Chawla and Hemant Kathuria 2019
H. Chawla and H. Kathuria, *Building Microservices Applications on Microsoft Azure*,
https://doi.org/10.1007/978-1-4842-4828-7_6

Logging the monitoring data is important for drawing performance trends over a period of months or years. Since AKS is a managed service, the monitoring components can be enabled with just a click, which saves days of effort in setting up monitoring. The following are features and tools used with AKS.

- Azure Monitor and Log Analytics

- Native Kubernetes Monitoring Dashboard

- Prometheus and Grafana

Azure Monitor and Log Analytics

Azure Monitor is a feature designed to monitor the performance of AKS clusters. It captures the performance data for

- Clusters

- Nodes

- Controllers

- Containers

Azure monitoring must be enabled manually to log all the information into a log analytics workspace. Log Analytics is the service in which all performance data can be logged. Let's see how this service is enabled while creating an AKS cluster.

Create an AKS Cluster from the Portal

Let's get started.

1. Sign in to the Azure portal.

2. Select Create a resource ➤ Kubernetes Service.

3. Enter the subscription resource group, Kubernetes
    cluster name, region, Kubernetes version, and DNS
    name prefix (as shown in Figure 6-1).

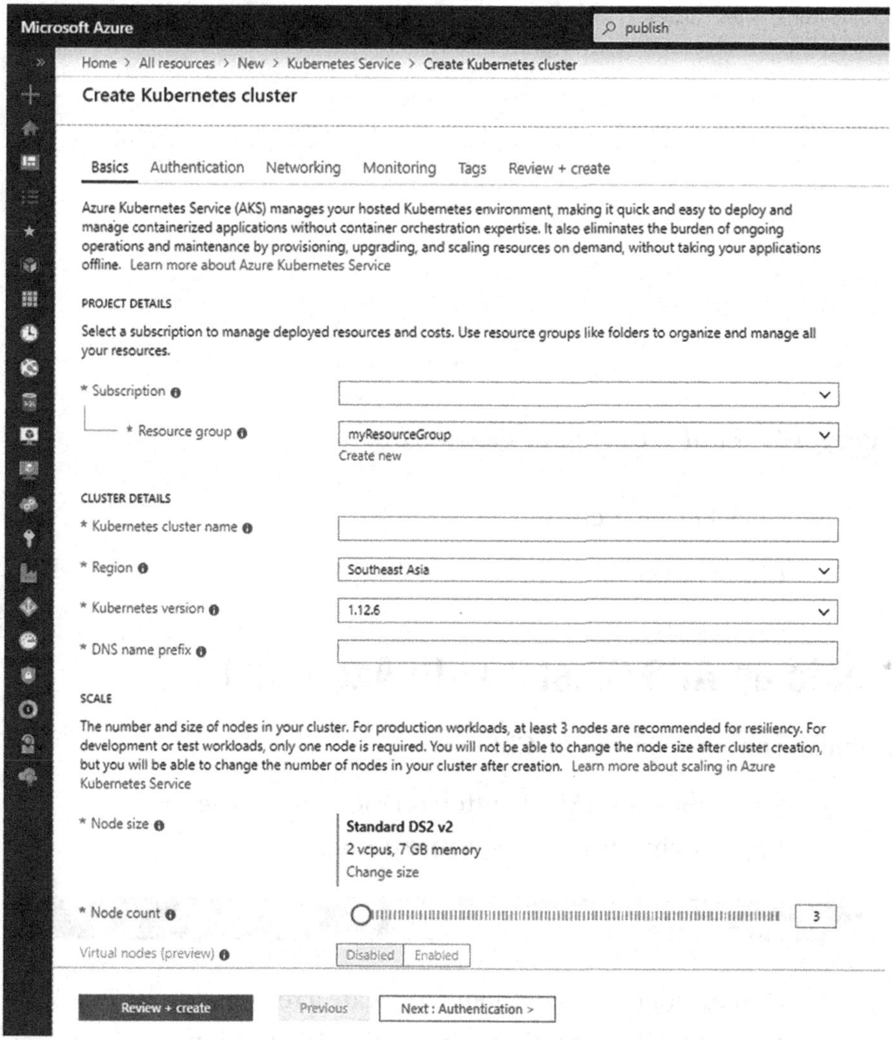

Figure 6-1. *Creating an AKS cluster on Azure portal*

4. Click Yes (as shown in Figure 6-2) to enable this
    option to set up Azure Monitor and Log Analytics for
    an AKS cluster.

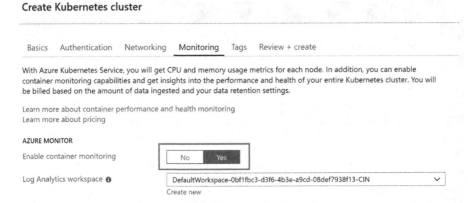

Create Kubernetes cluster

Basics   Authentication   Networking   **Monitoring**   Tags   Review + create

With Azure Kubernetes Service, you will get CPU and memory usage metrics for each node. In addition, you can enable
container monitoring capabilities and get insights into the performance and health of your entire Kubernetes cluster. You will
be billed based on the amount of data ingested and your data retention settings.

Learn more about container performance and health monitoring
Learn more about pricing

AZURE MONITOR
Enable container monitoring   [No   **Yes**]

Log Analytics workspace ⓘ   [DefaultWorkspace-0bf1fbc3-d3f6-4b3e-a9cd-08def7938f13-CIN   ∨]
                                    Create new

Figure 6-2. *Enable container monitoring*

5. Click Review + create.

6. Click Create.

Create an AKS Cluster with Azure CLI

Enable the option using Azure CLI.

1. Select the Cloud Shell button on the menu in the
    upper-right corner of the Azure portal.

2. Run the following command to create a resource group
    named myResourceGroup in the East US location.

```
az group create --name myResourceGroup --location
eastus
```

3. Once the resource group is created, running the
    following command creates an AKS cluster named
    myAKSCluster with one node.

```
az aks create \
    --resource-group myResourceGroup \
    --name myAKSCluster \
    --node-count 1 \
    --enable-addons monitoring \
    --generate-ssh-keys
```

Notice that we are enabling Azure Monitor for containers while creating using the -enabled-addons monitoring parameter.

Monitoring AKS Clusters

When the cluster is created either in the portal or in Azure CLI, go to the portal and type **Kubernetes** in the search box (as shown in Figure 6-3). Click the Kubernetes service that you have created.

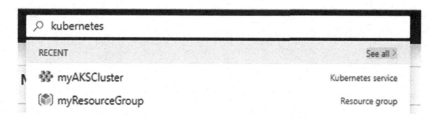

Figure 6-3. *Search AKS cluster*

There are two ways to monitor the AKS clusters: one is directly from an AKS cluster and other is to monitor all AKS clusters in the subscription.

Monitor from AKS

There are three monitoring options in AKS.

- Insights

- Metrics

- Logs

If the container monitoring option is not enabled, click Insights. The screen will look like the one shown in Figure 6-4.

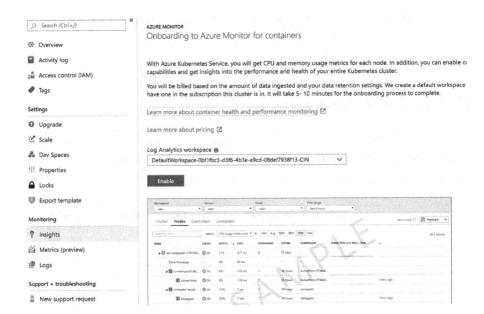

Figure 6-4. *Enable container monitoring*

Click Enable. Four AKS components can be monitored from the AKS cluster, as shown in Figure 6-5.

- Cluster

- Nodes

- Controllers

- Containers

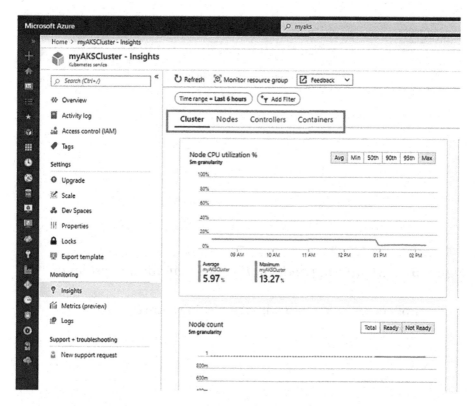

Figure 6-5. *AKS monitoring dashboard*

The performance chart displays four performance metrics, which are self-explanatory.

- Node CPU utilization

- Node memory utilization

- Node count

- Activity pod count

The Nodes tab is shown in Figure 6-6.

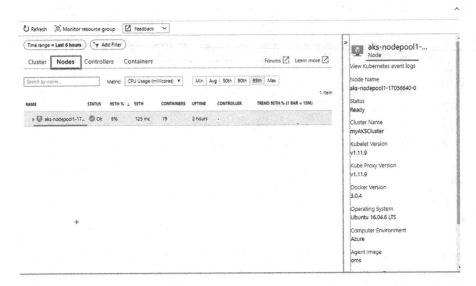

Figure 6-6. *Monitoring nodes in the AKS monitoring dashboard*

Controllers can be monitored, as shown in Figure 6-7.

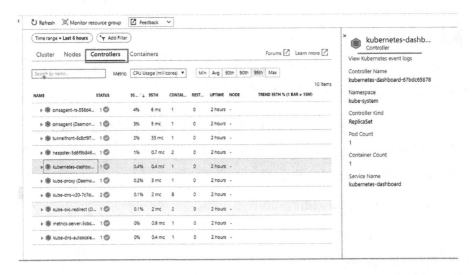

Figure 6-7. *Monitoring controllers in the AKS monitoring dashboard*

Similarly, containers can be monitored, as shown in Figure 6-8. In the Containers section, container logs and container live logs can be explored. Click **View container live logs**, as shown in Figure 6-8.

Figure 6-8. *Monitoring containers in AKS monitoring dashboard*

Figure 6-9 shows the container's performance data table from the log analytics workspace and live logging generated by the container engine to further assist in troubleshooting issues in real time.

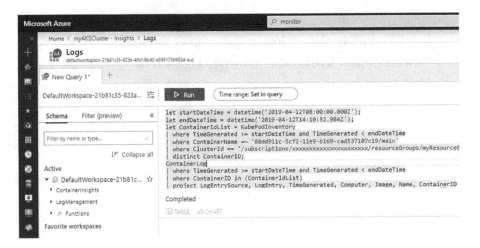

Figure 6-9. *Log analytics workspace*

Monitoring a Multi-Cluster from Azure Monitor

To view the health status of all deployed AKS clusters, select Monitor from the left-hand pane in the Azure portal. Under the Insights section, select Containers (as shown in Figure 6-10).

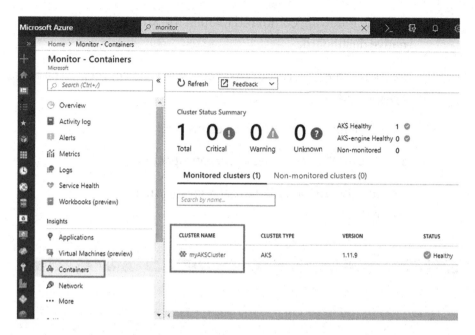

Figure 6-10. *Monitoring multiple AKS cluster*

From the list of clusters, select the cluster name to view the same monitoring information that was seen directly from AKS.

Native Kubernetes Dashboard

Kubernetes provides a dashboard as well. This dashboard offers comprehensive monitoring information. To open this dashboard, follow this process.

1. Select the Cloud Shell button on the menu in the upper-right corner of the Azure portal.

2. Run the following command to open the Kubernetes native dashboard.

```
az aks browse --resource-group myResourceGroup --name myAKSCluster
```

3. Open the following URL in a web browser.

```
127.0.0.1:8001/#!/overview?namespace=from
```

4. The dashboard looks like the screen shown in Figure 6-11.

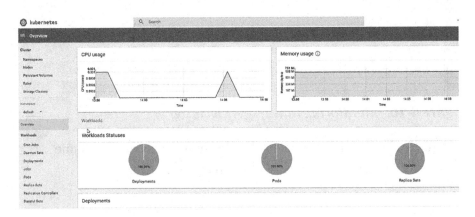

Figure 6-11. *Kubernetes dashboard*

It has all the components related to the Kubernetes ecosystem.

Prometheus and Grafana

Prometheus is an open source tool for monitoring and alerting. It captures comprehensive information about monitoring and can be connected to Grafana, another open source tool, for building dashboards. These are among the most commonly used tools to monitor Kubernetes. Therefore, it's worth mentioning these tools in this chapter. They connect seamlessly with AKS clusters. Figure 6-12 shows a Grafana dashboard.

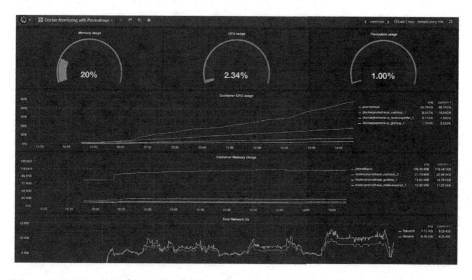

Figure 6-12. *Grafana dashboard*

Summary

In this chapter, you learned about various options for monitoring the AKS ecosystem. There are Azure, on-premise, and open source options that an administrator can choose from, depending on requirements. Azure Monitor and Log Analytics and the native Kubernetes dashboard are the easiest options to leverage. Prometheus and Grafana are popular monitoring/alerting and dashboarding tools, respectively.

CHAPTER 7

Securing Microservices

By the time you reach this chapter, you have a fair understanding of the microservices architecture and the common patterns used while implementing microservices, such as gateway aggregation, gateway routing, and gateway offloading. One of the most important aspects of implementing microservices is handling security. In an enterprise world, security is of utmost importance; various measures and audit processes are followed to make sure that data is secure and that services are not accessible to unauthorized users.

Although we briefly covered cross-cutting concerns in previous chapters, in this chapter, you will specifically learn about the various patterns and techniques available to secure your microservices. We will focus on how to integrate microservices with the leading identity provider, Azure Active Directory, which is almost a default choice for security implementation in an enterprise.

Authentication in Microservices

Authentication is the process of verifying a user's identity and making sure that only trusted users and clients can access the microservice. A commonly suggested practice in handling security is to use an API

© Harsh Chawla and Hemant Kathuria 2019
H. Chawla and H. Kathuria, *Building Microservices Applications on Microsoft Azure*,
https://doi.org/10.1007/978-1-4842-4828-7_7

gateway, as depicted in Figure 7-1. In this approach, the individual microservices cannot be reached directly and the traffic is redirected to individual APIs via a gateway once a successful authentication is performed.

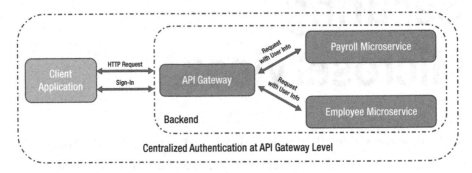

Figure 7-1. *Security at API gateway level*

Please note that in addition to authentication, an API gateway can also perform the following functionalities.

- Authorization
- Throttling
- Logging
- Response caching
- Service discovery
- IP whitelisting

In a scenario where services are exposed directly and without the intervention of an API gateway, a common technique is to use a dedicated security token service (STS). STS authenticates a user and then allows the user or client to access the API with the help of security cookies or an issued token, as depicted in Figure 7-2.

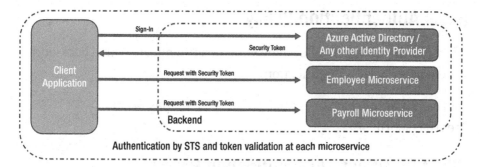

Figure 7-2. Authentication by STS

Implementing Security Using an API Gateway Pattern

While you can always develop a custom API gateway, here are two popular viable options that can be explored before implementing a custom API gateway.

Azure API Management

Azure API management is a fully managed service that allows customers to publish, monitor, transform, and secure APIs. With very little effort, it acts as a "front door" to your back-end APIs and provides functionalities like authentication and throttling.

Azure API management secures back-end APIs by using client certificates, tokens, IP filtering, and so forth.

Azure API management has a policies concept; you can use the following policies to secure your back end.

Basic Authentication Policy

A basic authentication policy allows you to authenticate with a back-end system using basic authentication.

```
<authentication-basic username="testuser" password=
"testpassword" />
```

Client Certificate Authentication Policy

A client certificate authentication policy allows you to authenticate with a back-end system by using a client certificate. The certificate first needs to be uploaded in API management, and then it can be identified by thumbprint in the authentication policy. The following sample policy allows you check if the request to the back-end API has the desired certificate thumbprint value or not.

```
<choose>
    <when condition="@(context.Request.Certificate == null
    || !context.Deployment.Certificates.Any(c => c.Value.
    Thumbprint == context.Request.Certificate.Thumbprint))" >
        <return-response>
            <set-status code="403" reason="Invalid client
            certificate" />
        </return-response>
    </when>
</choose>
```

JWT Validation Policy

If the back end is configured to be secured by JWT bearer tokens, Azure API management can preauthorize requests by using a validation JWT policy. An inbound policy can be added to validate the expiry and audience, and by signing the key of the passed token.

The following is an example token validation policy, which validates tokens issued by Azure Active Directory.

```
<validate-jwt header-name="Authorization" failed-validation-
httpcode="401" failed-validation-error-message="Unauthorized.
Access token is missing or invalid.">
    <openid-config url="https://login.microsoftonline.com/
    contoso.onmicrosoft.com/.well-known/openid-configuration" />
    <audiences>
        <audience>25eef6e4-c905-4a07-8eb4-0d08d5df8b3f
        </audience>
    </audiences>
    <required-claims>
        <claim name="id" match="all">
            <value>insert claim here</value>
        </claim>
    </required-claims>
</validate-jwt>
```

Ocelot

Ocelot is an open source, simple, lightweight, .NET Core–based API gateway that can be deployed along with your microservices. Ocelot is designed to work only with ASP.NET Core; basically, it is middleware that can be applied in a specific order. Ocelot is installed in an ASP.NET core project by adding Ocelot's NuGet package. Ocelot has many capabilities, but the following are the most popular.

- Routing

- Request aggregation

- Authentication/authorization

- Rate limiting

- Caching

- Logging

Hands-on Lab: Creating an Application Gateway Using Ocelot and Securing APIs with Azure AD

In this exercise, we will build an API gateway using Ocelot, which makes sure that the back-end API is protected by Azure AD OAuth Bearer Authentication. This exercise has three projects.

- **HRClientApp**. A desktop-based client app that allows the user to log in against Azure AD and then invokes the back-end employee service via APIGateway.

- **APIGateway**. An API gateway built using Ocelot and ASP.NET Core.

- **EmployeeService**. This represents the back-end API.

Setting up a Development Environment

Let's set up.

1. Install Visual Studio 2017.

2. Access the Azure portal to create application registrations.

Azure AD Application Registrations

Register an app in Azure AD to represent the back-end API.

1. Browse portal.azure.com and go to Azure Active Directory.

2. Click App Registrations (Preview).

3. Click New Registration.

4. Enter EmployeeServiceApi as the name and choose Accounts in Organizational Directory Only.

5. Click Register.

6. Click Expose an API in the left navigation bar.

7. Click Add a Scope and set an Application ID URI, as shown in Figure 7-3.

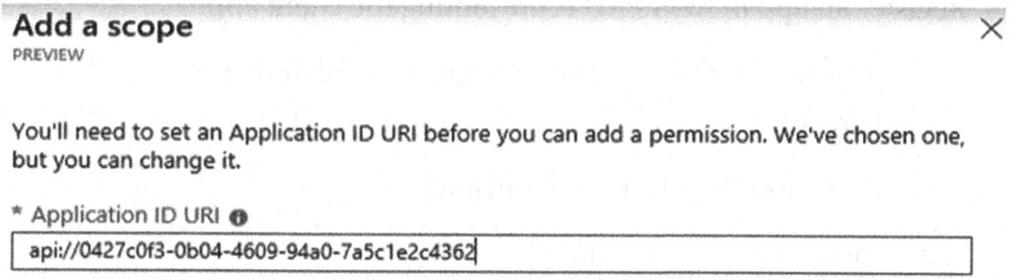

Add a scope   ✕
PREVIEW

You'll need to set an Application ID URI before you can add a permission. We've chosen one, but you can change it.

* Application ID URI ❶

api://0427c0f3-0b04-4609-94a0-7a5c1e2c4362

Figure 7-3. *Set Application ID URI*

8. Enter **allowaccess** as the scope name.

9. Enter **allowaccess** as the admin consent display name.

10. Enter **allow access** as the admin consent description.

11. Make a note of the Application ID/Client ID and Tenant ID; this will be needed in upcoming steps.

12. Take note of the complete scope name, as shown in Figure 7-4.

Figure 7-4. *Expose an API*

Register an app in Azure AD representing the client app.

1. Browse portal.azure.com and go to Azure Active Directory.

2. Click App Registrations (Preview).

3. Click New Registration.

4. Enter **HRClientApp** as the name and choose Accounts in Organizational Directory Only.

5. Click Register.

6. Once complete, click API Permissions in the left navigation bar.

7. Click Add Permission.

8. Click API My Organization Uses.

9. Search for EmployeeServiceAPI, as shown in Figure 7-5.

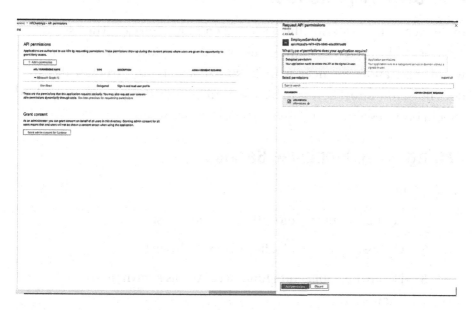

Figure 7-5. *Assign delegated permission*

10. Choose Delegated Permission, allowaccess scope, and click Add Permissions.

11. Click Authentication in the left navigation bar and add a redirect URI, as shown in Figure 7-6.

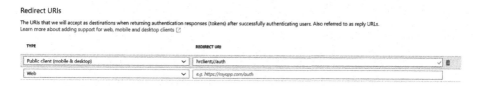

Figure 7-6. *Assign redirect URI*

12. Note the Application ID/Client ID, Tenant ID, and redirect URI; this will be needed later.

Develop an API Gateway, Back-end Service, and Client Application

In this section, we develop the back-end employee service. The user request will be routed from the gateway to the employee service. In a real-world application, you need to make sure that your infrastructure design does not allow direct access to back-end APIs. Only the gateway endpoints should be accessible to end users or client applications.

Setting up an Employee Service

Let's get started.

1. Launch Visual Studio 2017 as an administrator.

2. Create a project with File ➤ New ➤ Project.

3. Name the application Employee.Api, as shown in Figure 7-7.

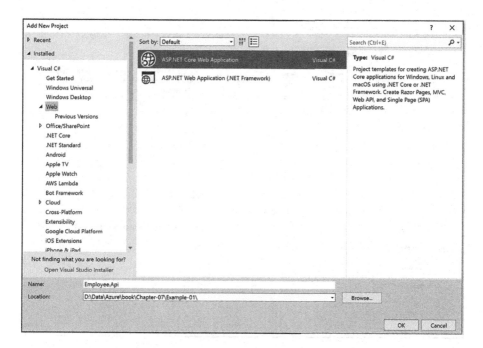

Figure 7-7. *Create employee API*

4. Click OK.

5. Choose the API and click OK. Select ASP.NET
    Core 2.2, as shown in Figure 7-8.

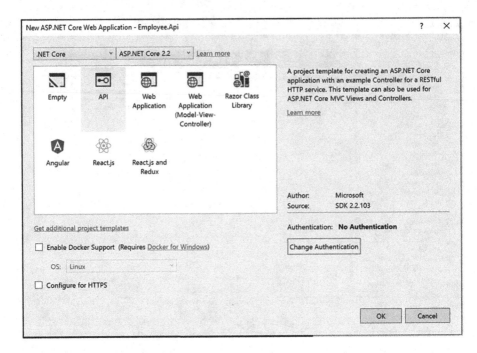

Figure 7-8. *Choose API template*

6. The default template adds the values controller.
    Keep it as it is; do not secure this controller. (We
    will use this controller to showcase the differences
    between secured and unsecured controller.)

7. Right-click the Controllers folder and add an empty
    class. Insert the following code. (We are keeping the
    controller very simple to showcase how to secure
    the controller against Azure AD and via the Ocelot
    API gateway.)

```
using Microsoft.AspNetCore.Authorization;
using Microsoft.AspNetCore.Mvc;
using System.Collections.Generic;
```

```
namespace Employee.Api.Controllers
{
    [Route("api/[controller]")]
    [ApiController]
    [Authorize]
    public class EmployeeController : ControllerBase
    {
        // GET api/values
        [HttpGet]
        public ActionResult<IEnumerable<string>> Get()
        {
            return new string[] { "Employee1", "Employee2" };
        }
    }
}
```

The Authorize attribute makes sure that the API can be accessed by an authenticated user or application only.

Now we will set up the Azure AD bearer token security for the project. Although the API gateway will perform the first level of authentication for us and will route the traffic only if the user is authenticated, we will need the bearer token middleware to create the user identity based on the passed token.

1. Right-click Dependencies under the project. Add the reference Microsoft.AspNetCore.Authentication. JwtBearer NuGet package.

2. Right-click the project and add a class name, AzureADOptions. This is required to represent a section in appsettings.json. Add the following code in the newly added class.

```
public class AzureADOptions
{
    public string Instance { get; set; }

    public string Domain { get; set; }

    public string TenantId { get; set; }

    public string ClientId { get; set; }
}
```

3. Open appsettings.json and add the section for Azure AD configuration.

```
"AzureAd": {
    "Instance": "https://login.microsoftonline.com/",
    "TenantId": "<<Tenant Id of Your Azure Ad>>",
    "ClientId": "<<Application / Client Id of Employee
    Service>>"
}
```

4. Open StartUp.cs and add the following code.

```
using Microsoft.AspNetCore.Authentication.JwtBearer;
using Microsoft.AspNetCore.Builder;
using Microsoft.AspNetCore.Hosting;
using Microsoft.AspNetCore.Mvc;
using Microsoft.Extensions.Configuration;
using Microsoft.Extensions.DependencyInjection;

namespace Employee.Api
{
    public class Startup
    {
```

```
public Startup(IConfiguration configuration)
{
    Configuration = configuration;
}

public IConfiguration Configuration { get; }

// This method gets called by the runtime. Use this
method to add services to the container.
public void ConfigureServices(IServiceCollection
services)
{
    AzureADOptions azureADoptions = new
    AzureADOptions();

    //"AzureAd" is the name of section in AppSettings.
    Config
    Configuration.Bind("AzureAd", azureADoptions);

    // Make sure the Name "AzureAdAuthenticationScheme"
    is same in Ocelot.json
    services.AddAuthentication(options =>
        {
            options.DefaultAuthenticateScheme =
            JwtBearerDefaults.AuthenticationScheme;
            options.DefaultChallengeScheme =
            JwtBearerDefaults.AuthenticationScheme; }
        )
        .AddJwtBearer(x =>
        {
            x.Authority = $"{azureADoptions.Instance}
            /{azureADoptions.TenantId}";
            x.RequireHttpsMetadata = false;
```

```
                x.TokenValidationParameters = new
                Microsoft.IdentityModel.Tokens.
                TokenValidationParameters()
                {
                    ValidAudience = azureADoptions.ClientId
                };
            });

        services.AddMvc().SetCompatibilityVersion
        (CompatibilityVersion.Version_2_2);
    }

    // This method gets called by the runtime. Use this
    method to configure the HTTP request pipeline.
    public void Configure(IApplicationBuilder app,
    IHostingEnvironment env)
    {
        if (env.IsDevelopment())
        {
            app.UseDeveloperExceptionPage();
        }

        app.UseAuthentication();

        app.UseMvc();
    }
  }
}
```

In this code, you are reading the Azure AD configuration from
appSettings.json and setting up the bearer token authentication.

If you run the Employee.Api project now, you will be able to browse the
API from the values controller, but you will get 401-unauthorized for the
API from EmployeeController.

Setting up an API Gateway

The following steps create an API gateway using Ocelot.

1. Right-click the solution and choose Add ➤ New Project. Select ASP.NET Core Application.

2. In the New Project dialog, choose Web ➤ ASP.NET Core Web Application, as shown in Figure 7-9.

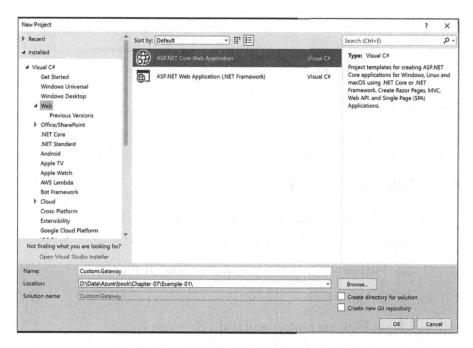

Figure 7-9. *ASP.NET Core web application*

3. Name the project Custom.Gateway. Click OK.

4. Select Empty. Choose ASP.NET Core 2.2 and click OK, as shown in Figure 7-10.

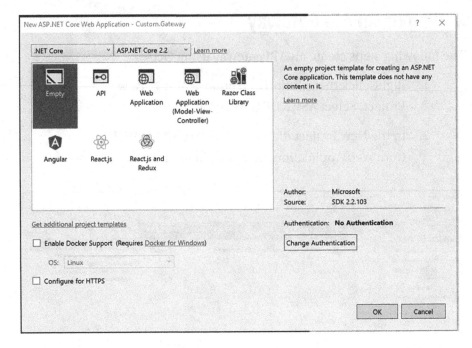

Figure 7-10. *Choose template*

5. Right-click Dependencies and add a reference for the following NuGet packages.

    a. Microsoft.AspNetCore.Authentication.JwtBearer

    b. Ocelot

6. Right-click the project and add a class name, AzureADOptions. This is required to represent a section in appsettings.json. Add the following code in the newly added class.

```
public class AzureADOptions
{
    public string Instance { get; set; }

    public string Domain { get; set; }
```

```
public string TenantId { get; set; }

public string ClientId { get; set; }
}
```

7. Open appsettings.json and add the section for Azure
    AD configuration.

```
"AzureAd": {
    "Instance": "https://login.microsoftonline.com/",
    "TenantId": "<<Tenant Id of Your Azure Ad>>",
    "ClientId": "<<Application / Client Id of Employee Service>>"
}
```

8. Open Program.cs and replace the content of the
    CreateWebHostBuilder method with the following
    content. (Here we are referring to a new file called
    ocelot.json, which we will create and configure in
    next step.)

```
public static IWebHostBuilder CreateWebHostBuilder(string[]
args) =>
            WebHost.CreateDefaultBuilder(args)
                .ConfigureAppConfiguration((hostingContext,
                config)=>{
                    config
                        .SetBasePath(hostingContext.
                        HostingEnvironment.ContentRootPath)
                        .AddJsonFile("appsettings.json", true,
                        true)
                        .AddJsonFile($"appsettings.
                        {hostingContext.HostingEnvironment.
                        EnvironmentName}.json", true, true)
```

```
                    .AddJsonFile("ocelot.json")
                    .AddEnvironmentVariables();
        })
        .UseStartup<Startup>();
```

9. Right-click the project and add a JSON file called
   ocelot.json. Add the following content to the newly
   created file. (Ocelot documentation is at https://
   ocelot.readthedocs.io/en/latest/.)

```
{
  "ReRoutes": [
    {
      "DownstreamPathTemplate": "/api/employee",
      "DownstreamScheme": "http",
      "DownstreamHostAndPorts": [
        {
          "Host": "localhost",
          "Port": 62550
        }
      ],
      "UpstreamPathTemplate": "/employee",
      "UpstreamHttpMethod": [ "Get" ],
      "AuthenticationOptions": {
        "AuthenticationProviderKey":
        "AzureAdAuthenticationScheme",
        "AllowedScopes": []
      }
    },
    {
      "DownstreamPathTemplate": "/api/values",
      "DownstreamScheme": "http",
```

```
    "DownstreamHostAndPorts": [
      {
        "Host": "localhost",
        "Port": 62550
      }
    ],
    "UpstreamPathTemplate": "/value",
    "UpstreamHttpMethod": [ "Get" ]
  }
],
"GlobalConfiguration": {
  "BaseUrl": "http://localhost:62420"
}
}
```

In this configuration, we are routing all the /employee route traffic to the downstream service at /api/employee route, and /value to /api/ values. Please make sure that you change the port values as per your environment. 62550 is the port on which the employee API is running. 62420 is the gateway project port.

Also, note that we specified AzureAdAuthenticationScheme as the authentication option for the employee route. Ocelot will execute the middleware associated with this scheme.

Now let's configure the scheme in Startup.cs. Open Startup.cs and add the following content.

```
using Microsoft.AspNetCore.Builder;
using Microsoft.AspNetCore.Hosting;
using Microsoft.Extensions.DependencyInjection;
using Ocelot.DependencyInjection;
using Ocelot.Middleware;
using Microsoft.Extensions.Configuration;
```

```
namespace Custom.Gateway
{
    public class Startup
    {
        public Startup(IConfiguration configuration)
        {
          ˚ Configuration = configuration;
        }
        public IConfiguration Configuration { get; }

        // This method gets called by the runtime. Use this
        method to add services to the container.
        // For more information on how to configure your
        application, visit https://go.microsoft.com/
        fwlink/?LinkID=398940
        public void ConfigureServices(IServiceCollection
        services)
        {
            AzureADOptions options = new AzureADOptions();

            //"AzureAd" is the name of section in AppSettings.
            Config
            Configuration.Bind("AzureAd", options);

            // Make sure the Name "AzureAdAuthenticationScheme"
            is same in Ocelot.json
            services.AddAuthentication()
                .AddJwtBearer("AzureAdAuthenticationScheme", x =>
                {
                    x.Authority = $"{options.Instance}/{options.
                    TenantId}";
                    x.RequireHttpsMetadata = false;
```

```
x.TokenValidationParameters = new
Microsoft.IdentityModel.Tokens.
TokenValidationParameters()
{
    //keep on adding the valid client ids of
    backend apis here.
    //If gateway has to support new services
    in future, add the client id of each
    backend api
    ValidAudiences = new[] { options.
    ClientId}
};
});

    services.AddOcelot();
}

// This method gets called by the runtime. Use this
method to configure the HTTP request pipeline.
public void Configure(IApplicationBuilder app,
IHostingEnvironment env)
{
    if (env.IsDevelopment())
    {
        app.UseDeveloperExceptionPage();
    }

    app.UseAuthentication();

    app.UseOcelot();
    }
    }
}
```

The added code is very much like the employee service, with a few differences. We added the Ocelot reference, and at the same time, we used the Audiences property instead of the Audience property.

Since it is a gateway and receives traffic that needs to be routed to multiple back-end services, we have to list the client IDs of all the supported back-end services. In this case, we have only one client ID available, hence adding only the same.

Now you can run both projects to browse the following URLs.

- `http://localhost:<<gatewayport>>/employee` returns 401 because it is secured against AAD.

- `http://localhost:<<gatewayport>>/value` returns 200 because it is an unsecured endpoint.

Setting up a WPF-based Client Application

Now that the back-end API and gateway host is up and running, we will create a client application where the user will perform the login, and the client application will invoke the API based on the token generated after login.

1. Right-click the solution, choose Add ➤ New Project, and select Windows Desktop.

2. Choose WPF App (.NET Framework). Name the application Custom.Client, as shown in Figure 7-11.

Figure 7-11. *Choose application type*

In this sample application, you will add two buttons: one to invoke employee service and one to invoke values service. Please make sure that you configure the APIGateway URLs, and not the direct service.

1. Right-click References and add the reference for the Microsoft.Identity.Client NuGet package.

2. Open MainWindow.Xaml and add the following code inside the Grid tag.

```
<Button Content="Invoke Employee API"
HorizontalAlignment="Left" Margin="71,60,0,0"
VerticalAlignment="Top" Width="667" Height="87"
Click="OnEmployeeClick"/>
<Button Content="Invoke Values API" HorizontalAlignment="Left"
Margin="71,186,0,0" VerticalAlignment="Top" Width="667"
Height="75" Click="OnValuesClick"/>
```

217

3.   Open MainWindow.Xaml.cs and add the following code.

```
public partial class MainWindow : Window
    {
        public PublicClientApplication _publicClientApp = null;

        //sample api scope : fccaa2fa-fe79-42fa-b8d5-
        e0ac6061ae99/allowaccess
        private string _apiScope = "API SCOPE VALUE FROM First
        Step";

        private string _apiGatewayEmployeeUrl =
        "http://localhost:<<your gateway port>>/employee";

        private string _apiGatewayValuesUrl =
        "http://localhost:<<your gateway port>>/value";

        public MainWindow()
        {
            InitializeComponent();

            string clientId = "Client id of <<HRClientApp>>
            application";

            string tenantId = "<<Your Tenant Id>>";

            //sample redirect uri: hrclient://auth
            string redirectUri = "Redirect Uri of
            <<HRClientApp>>";

            _publicClientApp = new PublicClientApplication(
            clientId, $"https://login.microsoftonline.com/
            {tenantId}");

            _publicClientApp.RedirectUri = redirectUri;

        }
```

```csharp
/// <summary>
/// Invokes Employees API Via API Gateway
/// </summary>
/// <param name="sender"></param>
/// <param name="e"></param>
private async void OnEmployeeClick(object sender,
RoutedEventArgs e)
{
    var authResult = await _publicClientApp.
    AcquireTokenAsync(new string[] { _apiScope
    }).ConfigureAwait(false);

    var employeResult = await GetHttpContentWithToken
    (_apiGatewayEmployeeUrl, authResult.AccessToken);

    MessageBox.Show($"Employee Result : {employeResult}");
}

/// <summary>
/// Invokes Values API Via API Gateway
/// </summary>
/// <param name="sender"></param>
/// <param name="e"></param>
private async void OnValuesClick(object sender,
RoutedEventArgs e)
{
    var valueResult = await GetHttpContentWithToken(_
    apiGatewayValuesUrl, string.Empty);

    MessageBox.Show($"Value Result : {valueResult}");
}
```

```csharp
/// <summary>
/// Makes an Http Call to API Gateway
/// </summary>
/// <param name="url"></param>
/// <param name="token"></param>
/// <returns></returns>
public async Task<string> GetHttpContentWithToken
(string url, string token)
{
    var httpClient = new System.Net.Http.HttpClient();
    System.Net.Http.HttpResponseMessage response;

    try
    {
        var request = new System.Net.Http.HttpRequest
        Message(System.Net.Http.HttpMethod.Get, url);

        if (!string.IsNullOrEmpty(token))
        {
            //Add the token in Authorization header
            request.Headers.Authorization = new System.
            Net.Http.Headers.AuthenticationHeaderValue
            ("Bearer", token);
        }

        response = await httpClient.SendAsync(request);

        response.EnsureSuccessStatusCode();

        var content = await response.Content.
        ReadAsStringAsync();

        return content;
    }
```

```
catch (Exception ex)
{
    return ex.ToString();
}
}
}
```

In this code, we utilized MSAL.NET to generate the bearer token and passed the same. MSAL.NET supports caching for optimized performance. Please visit https://github.com/AzureAD/microsoft-authentication-library-for-dotnet for online help.

1. Right-click the solution and click Set Startup Projects.

2. Select Start for all three projects, as shown in Figure 7-12.

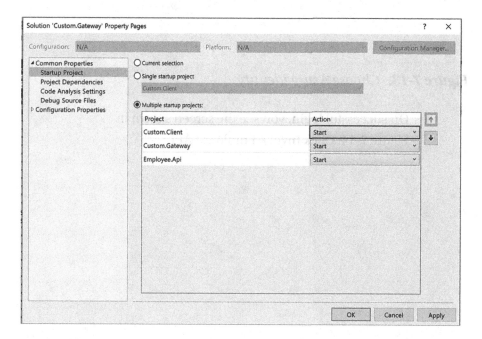

Figure 7-12. *Setting start up projects*

3. Click Invoke Employee API to launch the login
    screen from Azure Active Directory. Choose or enter
    the details of a valid user from your Azure Active
    Directory, as shown in Figure 7-13.

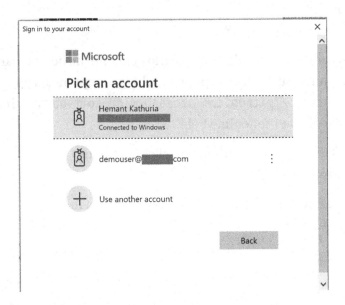

Figure 7-13. *Choose Azure identity*

4. On successful login, you see the screen shown in
    Figure 7-14. Click Invoke Employee API.

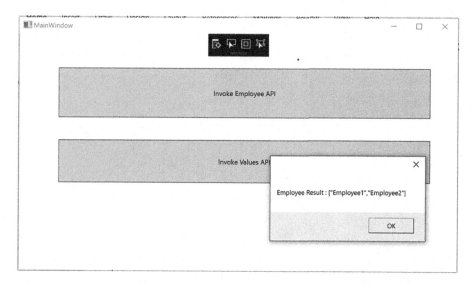

Figure 7-14. *Invoke Employee API*

Summary

In this chapter, you learned how to use an API gateway to implement security at the gateway level, instead of implementing it at an individual service. Although you can develop a custom API gateway, there are popular options available, like Azure API Management and Ocelot, which should be considered before developing a custom one.

CHAPTER 8

Database Design for Microservices

In this chapter, we will discuss the various critical factors of designing a database for microservices applications. As per the narrative of the microservices architecture, each microservice should have its own databases. Segregation based on data access helps fit the best technology to handle respective business problems. This means that a single application can use different database technologies, which is a concept called *polyglot persistence*. Let's look at important factors to consider when building a microservices application, as well as how the power of the Microsoft Azure platform can be harnessed to build highly scalable, agile, and resilient solutions.

Data Stores

Before starting the discussion on monolithic and microservices data architecture, it's important to understand the types of data stores available today. There are two major categories of data stores.

- RDBMS (Relational Database Management System)

- NoSQL (Not-only SQL)

© Harsh Chawla and Hemant Kathuria 2019
H. Chawla and H. Kathuria, *Building Microservices Applications on Microsoft Azure*,
https://doi.org/10.1007/978-1-4842-4828-7_8

RDBMS

RDBMS uses the ACID principle to run database operations. ACID principles are described as follows.

- **Atomicity**. All the changes in a transaction are either committed or rolled back.

- **Consistency**. All the data in a database is consistent all the time; none of the constraints will ever be violated.

- **Isolation**. Transaction data that is not yet committed can't be accessed outside of the transaction.

- **Durability**. Committed data, once saved on the database, is available even after the failure or restart of the database server.

Data consistency is natively built into RDBMS database solutions, and it's much easier to manage transactions with these solutions. However, scalability is the biggest challenge in RDBMS technologies.

- An RDBMS is designed to scale up/scale vertically (i.e., more compute can be added to the server rather than adding more servers). There are options to scale horizontally or scale out, but issues of write conflicts make it less scalable.

- An RDBMS is schema bound, and any change to the design needs a change in the schema. Therefore, an RDBMS can manage structured data very efficiently. However, there are limitations in handling semistructured or unstructured data.

- An RDBMS can be a big hassle for applications that change frequently; for example, shopping websites have multiple products with different features. In an

RDBMS, each feature acts as a column in a table, and that table has columns pertaining to all the features across products. Adding a product with new features means more columns need to be added to the table.

- An RDBMS fits better for applications in which data consistency and availability are highly critical (e.g., financial applications). However, for applications where data consistency requirements are relaxed and scalability is critical, NoSQL DB systems are a better fit.

NoSQL

NoSQL technology is based on distributed data stores that follow the CAP theorem to run database operations. The CAP theorem has three components.

- **Consistency**. Every read must receive the most recent write (i.e., every client reading data from the DB store should see the same data or an error).

  There are two extremes of consistency levels in CAP (i.e., strict (pessimistic) and eventual (optimistic)); however, there are some NoSQL DB stores that offer more consistency choices for flexibility.

  - **Strong (strict)**. All the reads are the most recent committed writes.

  - **Eventual**. There is no guarantee that reads are from the most committed writes; however, changes are eventually committed on all the replicas.

- **Availability**. Every request must receive a response without the guarantee of it being the most recent write.

- **Partition tolerance**. The system must continue to work during network failures between components. The system will continue to function even if one of the nodes fails.

The CAP theorem states that any distributed application can achieve any two functionalities, but not three, at the same time (see Figure 8-1).

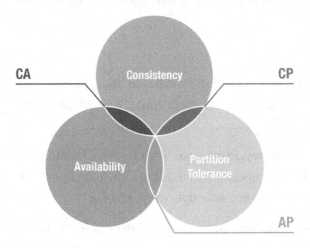

Figure 8-1. *CAP theorem*

Depending on the nature of the application, you can choose an intersection of two functionalities (i.e., consistency and partition tolerance, consistency and availability, or availability and partition tolerance). NoSQL technologies follow either availability and partition tolerance (e.g., Cassandra and riak) or partition tolerance and consistency (e.g., Hbase and MongoDB).

Alternative to ACID is BASE, which is followed by NoSQL data stores. BASE prefers availability over consistency; it is described as follows.

- **B**asically **A**vailable: The system appears to be available most of the time.

- **S**oft state: The version of data may not be consistent all the time.

- **E**ventual consistency: Writes across services are done over a period of time.

According to Dr. Eric Brewer, there is a continuum between ACID and BASE, and one can be closer to any end. Today's advanced applications are a careful mixture of both ACID and BASE solutions.

Monolithic Approach

Monolithic applications are the most preferred approach for small/medium-scale applications today. Monolithic applications use a single data store—either RDBMS or NoSQL.

Since RDBMS technology is well known and there is an adequate talent pool in the market, it is the first choice for the enterprises. This technology is relevant for only structured data, however. Applications like IOT, e-commerce, and social networking generate a lot of semistructured or unstructured data, which are classic use cases for NoSQL. A monolithic application doesn't have the flexibility to use both technologies in a single application. That's where the microservices architecture plays a key role and offers the flexibility to use polyglot persistence.

No matter which DB store is used, there are critical questions for an architect.

- How critical is the data?

- Do you prefer consistency over scalability, or scalability over availability?

- What type of data is it?

- Is the data structured, or semistructured, or unstructured?

For applications used in banking or finance, where data consistency and high availability is the utmost priority, RDBMS is the first choice. However, for applications like social networking, blogging, and gaming, where data availability and scalability is utmost priority, NoSQL fits best.

Microservices Approach

The microservices architecture approach suggests segregating an application into smaller and loosely coupled independent modules. This provides the flexibility to choose multiple database technologies for a single application. Both NoSQL and RDBMS technologies can be used in their respective strength areas. RDBMS is suggested for the modules where transactional consistency is critical and structured data is stored. However, NoSQL is suggested for modules where schema changes are frequent, maintaining transactional consistency is secondary, and semistructured or unstructured data is stored.

With the flexibility of choice, there are certain sets of challenges.

- Maintaining consistency for transactions spanning across microservices databases.

- Sharing, or making the master database records available across microservices databases.

- Making data available to reports that need data from multiple microservices databases.

- Allowing effective searches that get data from multiple microservices databases.

A microservices application's most critical challenge is the efficiency of transferring changes across the microservices. There are two possible approaches for this.

- Two-phase commit

- Eventual consistency

Two-Phase Commit

The two-phase commit approach is familiar to enterprises working with RDBMSs like SQL Server, MySQL, PostgreSQL, and Oracle. (NoSQL technologies don't natively support such transactions.)

Two-phase commit uses ACID principles to manage the transactions. As the name suggests, there are two phases.

- **Prepare phase**. A mediator called a transaction program manager helps complete transactions successfully. In the prepare phase, each program involved starts the transaction in their own database servers. Based on the status (successful or failed), the response is sent to the TP manager, which prepares the commit/rollback phase.

- **Commit/rollback phase**. In this phase, the TP manager instructs the programs to either commit or roll back. If the changes in the transactions were successful in all the participating programs, the TP manager instructs to commit the transactions; otherwise, it instructs a rollback.

In on-premises scenarios, the Microsoft Distributed Transaction Coordinator (MSDTC) on Windows Server environments is used. Even though MSDTC is not available on PaaS database services in Azure, distributed transactions are still achievable in Azure SQL DB managed instances.

There are certain limitations in two-phase commit transactions that make it a less feasible option for microservice applications.

- Since a two-phase commit between RDBMS and NoSQL technologies is not natively available, and building a new framework would be a huge effort, architects are restricted on the choice of technology for an RDBMS.

- Scalability is a concern because transactions can be blocked or may need more resources on the participating database instances. In the event of blocking, there may be a delay in the commit or rollback. This situation must be carefully handled in the application because it can make the entire operation halt or have long waits.

- If the transaction spans across multiple microservices, the performance of the DB operations can be highly inconsistent during peak loads.

- It's recommended to use the API layer for any database operations of a microservices application.

If a transaction spans across multiple services, it's important to make the changes to all the services asynchronously. If the changes are synchronous, it brings the entire operation to a halt until the transaction is completed. These delays are unaffordable for the majority of applications; therefore, eventual consistency is most suitable for such transactions.

Eventual Consistency

Eventual consistency is the recommended way to manage transactions for any distributed application. It follows BASE principles, which states that data is replicated across microservices over time, and that this operation is asynchronous in nature.

Let's discuss the following to better understand.

- How do we maintain the consistency of critical finance data?

- Can Microservices applications work for applications that need high consistency?

The important point to remember is that microservices applications provide the option to use the best technology to handle the business problem. In a **H**uman **R**esource **M**anagement **S**ystem, RDBMS can manage finance data, where ACID principles can be used to ensure consistency. However, to store employee information like name, address, and other attributes, a document store like Cosmos DB can be used. Maintaining strict consistency is not possible in microservices applications due to disparate data stores. Therefore, eventual consistency is recommended to update the changes efficiently across services.

Since communication is asynchronous, and data technologies can be disparate with different syntax and semantics, to make replication feasible, queuing technologies like RabbitMQ, NServiceBus, and MassTransit, or even a scalable bus in cloud, are used.

Harnessing Cloud Computing

It's important to understand the various database options available on Microsoft Azure to reinforce the microservices ecosystem. As you know, this is the era of cloud computing, and the focus of organizations has shifted from on-premises data centers to cloud computing. Companies like Microsoft, Amazon, Google, and Alibaba have greatly invested in cloud platforms. Microsoft Azure has various database options on IaaS and PaaS, which can build highly scalable and resilient applications.

Figure 8-2 illustrates the following ways to host a database service on Microsoft Azure.

- **Infrastructure as a Service**. A database instance on a virtual machine

- **Platform as a Service**. Managed instances and Azure SQL DB

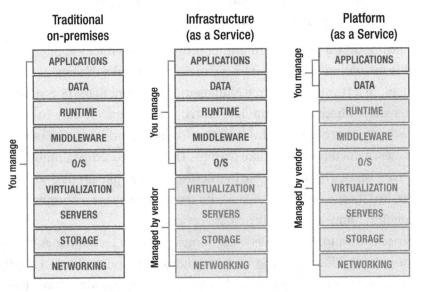

Figure 8-2. *Deployment models on Microsoft Azure*

Infrastructure as a Service (IaaS)

IaaS is typically managed in the same way as your own private data center. SQL Server—or any RDBMS or NoSQL database instances—can be installed on a virtual machine (VM). Configuration, backups, high availability, and disaster recovery must be done by administrators.

In Figure 8-2, the end user needs to manage everything above the virtualization layer. Virtual machines are backed by uptime SLAs (service-level agreements). Based on the criticalness of the application and acceptable downtime, VMs need to be configured (see Table 8-1).

Table 8-1. *SLAs of VMs (subject to change)*

Deployment Type	SLA	Downtime per week	Downtime per Month	Downtime per Year
Single VM	99.90%	10.1 minutes	43.2 minutes	8.76 hours
VM in Availability Set	99.95%	5 minutes	21.6 minutes	4.38 hours
VM in Availability Zone	99.99%	1.01 minutes	4.32 minutes	52.56 minutes

In regards to a SQL Server instance's high availability, virtual machines should be in the availability set. An availability set is the logical grouping of virtual machines allocated in an Azure data center; it ensures high availability of resources in the event of hardware or software failure in the same data center. To improve the uptime, SQL server AlwaysOn availability sets can be configured. it can replicate the data in both synchrous and asynchronous fashion. It's the native SQL server option for HADR.

Platform as a Service (PaaS)

With PaaS, the entire infrastructure is managed by cloud service providers. As shown in Figure 8-2, administrators need to manage their applications and data; therefore, managing resources becomes convenient for administrators. Let's take an example of Azure SQL DB; the entire OS layer and SQL Server layer, including version upgrades, are managed by Microsoft. End users need to manage the databases, respective application schemas, and T-SQL code. That's why it's also called Database as a Service.

Azure SQL Database and managed instances for SQL Server, PostgreSQL, MySQL, and MariaDB are PaaS options on Microsoft Azure. With Azure SQL

DB, there are few limitations in terms of functionalities, like cross-database transactions, CLR, collation changes, backups/restore, SQL Server agents, database mail, and so forth. Due to these limitations, on-prem database migration to Azure need changes at the application and functionality level.

To give more flexibility to the end user, Microsoft launched managed instances options for SQL Server, MariaDB, PostgreSQL, and MySQL. Managed instances have a surface area that is nearly 100% compatible to on-prem SQL Server, and yet it's a PaaS service. Migration to these instances is seamless, and there is no major change required for applications. We discuss both options in detail in the coming sections of this chapter.

Database Options on Microsoft Azure

Let's discuss some of the database options on Microsoft Azure that can be harnessed for microservices applications.

- RDBMS databases

- NoSQL databases

RDBMS Databases

For IaaS, any database software can be installed on VMs in Azure, and it can be managed by the DBA. Most of the prevalent solutions today are supported on Azure IaaS. However, PaaS services for databases are available in two options.

- Database throughput unit (DTU)–based purchasing model

- vCORE-based purchasing model

Azure SQL database follows a DTU purchasing model; however, SQL managed instances follow vCORE-based purchasing models. Let's briefly discuss Azure SQL DB and a SQL Server managed instance.

Note   DTU is the blend of CPU, memory, and IO resources to support database workloads. As a conventional approach, we are used to thinking in terms of memory, CPU, and IO.

Azure SQL DB

This is the first database PaaS service on Microsoft Azure. The resource allocation is done based on the DTU. Understanding the DTU requirement can be tricky when planning a database deployment. Autoscaling of resources can be helpful when the compute is undersized. The flexibility to autoscale helps to optimize resources. DBAs can start with lesser resources and scale up or down based on user demand.

Table 8-2 describes the resource allocation per the DB tier.

Table 8-2. Comparison of Basic, Standard, and Premium SQL Databases

	Basic	Standard	Premium
Target workload	Development and production	Development and production	Development and production
Uptime SLA	99.99%	99.99%	99.99%
Backup retention	7 days	35 days	35 days
CPU	Low	Low, Medium, High	Medium, High

(continued)

Table 8-2. (*continued*)

	Basic	Standard	Premium
IO throughput (approximate)	2.5 IOPS per DTU	2.5 IOPS per DTU	48 IOPS per DTU
IO latency (approximate)	5 ms (read), 10 ms (write)	5 ms (read), 10 ms (write)	2 ms (read/write)
Columnstore indexing	N/A	S3 and above	Supported
In-memory OLTP	N/A	N/A	Supported
Maximum storage size	2 GB	1 TB	4 TB
Maximum DTUs	5	3000	4000

These values may change; refer to MSDN for the latest information.

When autoscaling to higher tiers, minimal interruption in service may occur because autoscaling causes the creation of new compute instances for a database, followed by switch routing of connections to the new compute instance. Moreover, there is an option of elastic pool where there can be group databases on a single Azure SQL DB instance. This option is used for SaaS applications, where a database resource can be allocated to each customer. Since multiple databases share a pool of resources, the usage can be highly optimized when the load is unpredictable for each customer. If one customer is using less compute, and another customer is using more, there is a balance to manage within the allocated resources; otherwise, every customer database will be equally sized, which may end up causing resource overprovisioning.

SQL Managed Instance

SQL Managed Instance is a database PaaS service that has nearly 100% of the surface area of a SQL Server instance. Unlike Azure SQL DB, migration to a managed SQL instance is done without making any application code changes. It falls under the vCORE-based purchasing model. There are two tiers for this service.

- General purpose

- Business critical

For an Azure SQL database, autoscaling is based on DTUs; however, for managed instances compute, storage and IO resources can independently scale. Managed instances are available for the following relational data stores as well.

- MariaDB

- PostgreSQL

- MySQL

There are three tiers for these database services. A resources comparison is shown in Table 8-3.

- **Basic**. This is used for small-scale applications.

- **General Purpose**. This is used for applications that need balanced compute and memory with scalable I/O throughput.

- **Memory Optimized**. This is used for applications that need in-memory performance for faster transaction processing.

Table 8-3. *Comparison of Database Tiers*

	Basic	General Purpose	Memory Optimized
Compute generation	Gen 5	Gen 5	Gen 5
vCores	1, 2	2, 4, 8, 16, 32	2, 4, 8, 16
Memory per vCore	2 GB	5 GB	10 GB
Storage size	5 GB to 1 TB	5 GB to 4 TB	5 GB to 4 TB
Storage type	Azure Standard Storage	Azure Premium Storage	Azure Premium Storage
Database backup retention period	7 to 35 days	7 to 35 days	7 to 35 days

These values may change; refer to MSDN for the latest information.

NoSQL Databases

Microsoft Azure has a PaaS solution called Cosmos DB that supports all four types of NoSQL data.

- **MongoDB** document store
- **Cassandra** column family store
- **Gremlin** graph store
- **Table** key/value pair

The ARS (atoms, records, sequence) model maps to different data models very easily. It's a globally distributed and multimodel database service that can scale elastically or independently across Azure geographic regions. A request unit (RU) is the unit of throughput on Cosmos DB; each operation (e.g., read/write/store) procedure has a deterministic RU value. Another Cosmos DB advantage is the range of programming models that it offers.

- MongoDB
- Cassandra

- Gremlin

- Table

- SQL

It's very easy for a programmer to work with Cosmos DB because there's just a change in the connection string (with minimal code changes) to migrate your old application. You can choose any programming model from the preceding list to build your application on Cosmos DB.

Overcoming Application Development Challenges

By now, you should have clarity on the database services available on Microsoft Azure. Let's discuss how enterprises are harnessing these capabilities to overcome challenges during building a microservices application. Many enterprises use the capabilities of both the monolithic and the microservices architecture. Since data is the most critical and complex part of architecture, enterprises choose to modularize the web and app tiers and maintain a monolithic database centrally. By maintaining the data centrally, the following challenges are resolved by default.

- The ACID principles for transactions spanning across a microservices database.

- Sharing, or making the master database records available across microservices databases.

- Making data available to reports that need data from multiple microservices databases.

- Making an effective search that gets data from multiple microservices databases.

This approach can work very well for small and medium-sized applications. If the application is large and mission critical, the microservices architecture is recommended.

Let's revisit all the major challenges for a microservices application (in terms of databases) to better understand how Microsoft Azure can help you overcome these challenges.

Challenge 1

Maintain consistency in transactions spanning across microservices databases.

Resolution

The data changes that span across microservices should be asynchronous. Eventual consistency should be followed to transfer the changes.

In a failed operation, rolling back data changes to all the databases may be needed. It's not easy to roll back across distributed data stores; compensating transactions are the way to manage such operations.

Challenge 2

Share or make the master database records available across microservices databases.

Resolution

One service must maintain the master data, and then replicate the data and changes to other databases asynchronously.

Challenge 3

Make data available to reports that need data from multiple microservices databases.

Resolution

First, command and query resource segregation (CQRS) must be implemented. CQRS is an architecture pattern to make applications more scalable. It is recommended to write application code to separate command and query requests.

- **Command** (create, update, delete)
- **Query** (read)

This segregation offers the flexibility to choose from different data sources. Queries can be easily routed to the caching layer or read-only copy of the database. In Azure SQL DB, read-only connections can be routed to a secondary replica if an active sync replication is configured. Commands (i.e., write operations) can be routed to the primary database. This helps to reduce the load on the primary replica and the issue of resource crunch or blocking on the primary database, due to read queries.

If microservices database reports are slow due to data size, a data warehouse solution should be implemented. In Microsoft Azure, a solution called Azure SQL Data Warehouse manages structured data well. For unstructured data, or a mix of both structured and unstructured data, a solution called Azure Databricks can be used. All reporting or read-only queries can be routed to separate data warehousing systems.

If the data set is small, and there were no major changes on the source databases, even indexed views in SQL Server can help solve this challenge.

Challenge 4

Allow effective searches that get data from multiple microservices databases.

Resolution

This challenge should be handled at the application layer through API composition or through a caching layer.

Summary

This chapter discussed the various design patterns to manage data for microservices applications. There is a lot of conceptual information that is helpful in understanding the rationale behind choosing specific technologies. All the major options on Microsoft Azure can reinforce the impact of microservices applications with advanced cloud platform functionalities.

CHAPTER 9

Building Microservices Applications on Azure Stack

If we look back at how the data centers of organizations have evolved over the years, we realize a trend. Technological advancements in computing hardware and software are aimed at maximizing the use of the resources available. Today, we talk about hyperscale, with organizations adopting public cloud platforms to deliver all sorts of computing resources and cutting-edge technology platforms like IoT, blockchain, and machine learning services, and so forth. This enables organizations to shift their focus from operating and maintaining their data centers to addressing their business challenges using a public cloud.

It's important to realize that the cloud isn't just a location; it's a model that can be implemented in multiple ways. Although public cloud platforms have a lot to offer, there are still a lot of scenarios in which organizations can't afford to run certain workloads in the public cloud and wish to run these specific workloads within their premises. Reasons for this may include mission-critical workloads that are disconnected from the

© Harsh Chawla and Hemant Kathuria 2019
H. Chawla and H. Kathuria, *Building Microservices Applications on Microsoft Azure*,
https://doi.org/10.1007/978-1-4842-4828-7_9

Internet, or complying with the country's/organization's data residency regulations.

By embracing a hybrid cloud strategy, an organization can seamlessly manage on-premise and public cloud workloads through a single interface. You can run different workloads on-premise and use the public cloud when needed. In this chapter, we discuss a hybrid cloud solution called Microsoft Azure Stack and how it can be utilized for microservices applications.

Azure Stack

Azure Stack is Microsoft's implementation of this robust definition of the cloud in a hybrid mode. It's an extension of Azure that empowers your data center to provide a subset of Azure services. Microsoft has built this integrated solution with hardware OEMs (e.g., HPE, Dell, Lenovo, and Cisco) by putting Azure layer on top of it. This box can connect to your local infrastructure easily. It gives you the flexibility to build an application on Azure Stack and extend it to the Azure cloud when needed.

Services Available in Azure Stack

Azure Stack offers various services for your local data centers; however, in context to Microsoves application, it can offer

- Infrastructure as a Service (IaaS)
  - Virtual machines
  - Storage
  - Network

- Platform as a Service (PaaS)

  - Azure App Services

  - Azure Functions

  - Container orchestrators: Azure Service Fabric and Azure Kubernetes Services

  - Databases: SQL Server and MySQL

If services are not available natively in Azure Stack, there is an option to leverage Azure Marketplace to bring the solution to the box. It saves months/years of efforts to build such capabilities on-premises from scratch and makes these services natively available on Azure Stack.

Here's a cursory list of the inherent benefits a hybrid cloud implementation brings to the table.

- Enhances developers' productivity by giving them access to advanced and agile development platforms of their choice in their own data centers.

- Gives organizations the flexibility to rebalance their workloads as their requirements change by easily moving applications and data at will between Azure Stack and Azure.

- Enables enterprises to create their own private clouds with access to cutting-edge cloud services, while keeping their code and data within the bounds of their own firewalls.

Azure Stack Deployment Modes

The deployment mode is Azure Stack's most critical factor because it decides to set of features available to be offered from Azure Stack. There are two modes of Azure Stack deployment.

- **Connected**. Connected to the Azure public cloud through the Internet or MPLS

- **Disconnected**. Disconnected from the Azure public cloud

The true power of Azure Stack can be harnessed in a connected scenario because the capability to move a workload from Azure Stack to Azure, or vice versa, can be achieved natively. Identity management, registration, syndication, and patching can be easily managed by connecting directly to an Azure subscription over the Internet; however, for a disconnected Azure Stack setup, the hybrid scenario of moving the workload from Azure Stack to Azure (or vice versa) is not available, and all the basic functions must be manually done.

Now it's time to discover how various services on IaaS/PaaS/SaaS can be offered from this platform.

Offering IaaS

The most fundamental demand of all cloud tenants is to have highly available, configurable, and scalable IaaS offerings. Azure Stack delivers IaaS offerings with a user experience identical to that on Azure. Let's consider the example scenario shown in Figure 9-1, with two tenants using Azure Stack's IaaS offerings for different applications.

Tenant A has deployed a web application that is running on three VMs that rely on a load balancer to expose its public IP, and then distributes app user requests between the three VMs. For storage, tenant A has deployed a VM that runs SQL Server (or any other database system). Apart from this web application, tenant A has also deployed a dev/test environment running on a different virtual network that is isolated from the production environment. Tenant B has simply deployed a SharePoint farm running in three VMs connected by a virtual network.

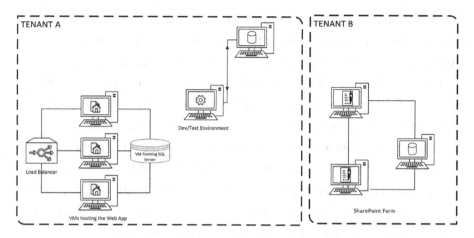

Figure 9-1. *IaaS scenario*

Since they are different tenants, neither consumer can access the cloud resources used by the other. The Azure Resource Manager (ARM) plays a key role in ensuring a consistent experience between Azure Stack and Azure by delivering identical REST APIs. ARM lets you realize the true potential of a hybrid cloud deployment by letting users automate the mechanics of their deployment and easily move workloads between Azure Stack and Azure. To showcase how a hybrid model is relatable to enterprise data centers today, let's revise the Figure 9-1 example. Tenant A may find moving the dev/test environment to Azure more prudent since it frees up resources if there is a hike in demand in the production environment. Tenant B can leverage ARM templates to automate compute scaling; for example, as the database content increases.

PaaS On-Premises Simplified

In a world where newer technologies that incorporate things like artificial intelligence, machine learning, and natural language processing are disrupting industries, it is more prudent for enterprises to channel a lot of their resources into innovation and development. Although IaaS and virtualized hardware are important, PaaS offers the perfect playground to innovate and deploy applications faster because you don't have to concern yourself with the virtualized hardware that the application runs on.

Let's visit the example shown in Figure 9-2 to see how building and deploying the same application is much simpler using PaaS. Tenant A had to explicitly create, configure, and manage VMs and define virtual networks and load balancers for the production and testing environments. If tenant A used PaaS offerings, the same traditional web application could be deployed using Azure App Service. Tenant A only needs to provide the application code and then let Azure App Service automatically deploy it across the necessary compute and networking infrastructures.

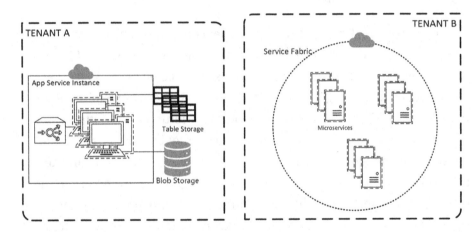

Figure 9-2. *PaaS scenario*

Tenant B can deploy the SharePoint farm using Service Fabric, where the architecture of the application is based on microservices instead of an n-tier model. Service Fabric, like App Service, handles the underlying infrastructure by hosting the application automatically, and managing the microservices deployed.

Azure Stack offers PaaS services like these out of the box, providing an easier way to fulfil a cloud platform's real goal, which is to support and run applications.

Another powerful example of a PaaS offering is in an IoT scenario where there's a need for low-latency communication between a device and the code controlling it. Services like IoT Hub let you create and run the control code locally and manage communication with the device. The emergence of PaaS offerings also makes it easier to use the arrays of other managed services, like Active Directory, blobs, tables, storage options, and so forth.

The best part is that the out-of-the-box PaaS offerings available on Azure Stack are just the tip of the iceberg. Access to Marketplace lets you download a plethora of other PaaS features delivered by third-party vendors such as DataStax, Cassandra, and Barracuda, as well as open source offerings from Linux distributions like Red Hat, Canonical, and so forth.

SaaS on Azure Stack

A huge challenge for software companies that create SaaS (Software as a Service) applications is that they have customers that demand on-premises versions. Azure Stack can let you build and deploy an application both on-premises and on Azure, which eliminates the problem of building different versions of the application to be run on-premises and on the cloud. SaaS applications need to be highly scalable, reliable, and easy to update because it makes sense to create SaaS applications using microservices and Service Fabric on Azure, along with an ARM template to automate the deployment of the application. So, if there is a demand for an on-premises version of the application, you can seamlessly move the software and the ARM template to Service Fabric running on Azure Stack.

While we understand the value proposition of Azure Stack, it's important to also understand how microservices applications are deployed on this solution. Taking reference from Chapters 1 and 2, microservices applications need the following components.

- Orchestrators

- Containers

- Stateful/stateless services and guest executables

- API gateway

- Databases

- Security

- Monitoring

Let's get into the details of the how Azure Stack is set up to be ready for the infrastructure deployment. There are three major steps to set up the Azure Stack.

1. **Azure Stack registration**. When the Azure Stack box is set up, the first step is to register with your Azure subscription to enable the inflow of services to Azure Stack. The Azure Stack portal is shown in Figure 9-3. The portal may change as new updates are released.

Figure 9-3. *Azure Stack portal*

2. **Marketplace syndication.** After registration, the next step is Azure Marketplace syndication, which brings IaaS and PaaS images to Azure Stack. In connected mode, this is seamless; however, in disconnected mode, all the required images are first downloaded to the Internet-connected machine and then moved to Azure Stack. During Marketplace integration, Windows Server (with containers) can be downloaded to build a container ecosystem.

3. **ADFS integration.** To integrate Azure Stack with your data center, you could either connect it with Azure Active Directory or use Active Directory Federation Services (ADFS) to connect on-premise AD to Azure Stack. Post integration, role-based access can be implemented and multiple users/ roles can be accessed or modified.

Now, Azure Stack is ready for use. Let's briefly discuss how a microservices ecosystem can be built on Azure Stack.

- **Orchestrators.** Two orchestrators are natively available on Azure Stack: Service Fabric and Azure Kubernetes Services (AKS). Other orchestrators, like OpenShift, Docker Swarm, and Mesos DC/OS can be configured on Azure Stack virtual machines. Both Service Fabric and AKS are currently under preview and will be generally available on Azure Stack as per the timelines by Microsoft product teams.

- **Containers.** Containers are the first options for hosting microservices applications. It can be easily built on the Azure Stack environment using IaaS, or you can directly build container images in Visual Studio to deploy it on AKS or Service Fabric clusters on Azure Stack.

- **Stateful/stateless services and guest executables.**
  Apart from containers, the Service Fabric orchestrator
  offers two other ways to build microservices workloads:
  stateful/stateless services and guest executables.
  Service Fabric is natively available on Azure Stack, and
  there is no change in the experience with Azure public
  cloud. It is convenient for a developer to build, test,
  and deploy microservices application efficiently. This is
  covered in Chapter 9.

- **API Gateway.** There is no native API management
  available on Azure Stack; however, an open source
  API management called Ocelot can be configured on
  Azure Stack virtual machines. Configuration is shared
  in Chapter 7.

- **Databases**. Azure Stack offers SQL Server and MySQL
  resource providers DB PaaS services. Other databases,
  like PostgreSQL, Oracle, MongoDB, and so forth, can
  be deployed on Azure Stack virtual machines.

- **Security**. Security can be offloaded to an API gateway
  solution, which can perform the following tasks:

  - Authentication and authorization

  - Throttling

  - Logging

  - Response caching

  - Service discovery

  - IP whitelisting

If needed, Azure Active Directory or ADFS can be utilized to implement role-based access for applications.

- **Monitoring.** There is no native solution on Azure Stack for monitoring; however, Azure Application Insights can be used, or a solution from Azure Marketplace can be implemented to manage monitoring for microservices applications.

Summary

To conclude this chapter, let's look at what Azure Stack means for an application designed with a microservices-oriented architecture deployed over clusters of containers. Such applications running on Azure Stack could help organizations take advantage of the low latency and meeting of compliance requirements by running relevant services of their application hosted on-premises.

At the same time, the application can leverage more compute-intensive services, such as machine learning models using the public cloud's ample resources, while leaving plenty of capacity to increase the number of instances running on-premises. Azure Stack and a hybrid cloud strategy give organizations access to a plethora of PaaS offerings that can be deployed on-premises and on Azure, seamlessly accelerating organizations to become cloud-ready on their own terms.

Azure Stack is a solution that can be leveraged to implement all three forms of cloud computing (IaaS, PaaS, and SaaS), which gives organizations the cloud in a box.

Index

A

ACID transactions, 32
Active Directory Federation
    Services (ADFS), 253
Active secondary replicas, 51
Agile methodology, 7, 15
Agility, 3, 7–8
AKS clusters, monitoring
    dashboard, 185–187
    enable components, 184
    multi-cluster, 188, 189
    options, 184
    performance metrics, 185
AMQP protocol, 30
API gateway, 198
    aggregation, 23–25
    Azure pattern, 28
    entry point, 23
    offloading, 27, 28
    routing, 25, 26
    single custom, 23
Application architecture, 5
Application Insights, 112
    registration process, 121
    SDK installation, 122
    view, 137

Application monitoring
    Application Insights, 113
    ASP.NET Core (*see* ASP.NET
        Core creation)
Application performance
    management (APM)
    service, 112
ASP.NET Core
    debug the application, 83, 84
    EmployeeDataAPI, 76–79,
        81, 82
    reliable collections, 58–63, 65–67
    reverse proxy, 67–69, 71, 73, 74,
        76–79, 81, 82
    Service Fabric status, 58
    setting up development
        environment, 57, 58
    stateless, 116
ASP.NET Core creation
    Application
        Insights, 118–119, 137
    ASP.NET Core 2.2, 117
    Azure subscription, 120
    EmployeeController, 123–132
    grouped results, 138
    registration, 121
    SDK, 121–122

Printed in the United States
By Bookmasters